T0358382

Nonstandard Finite Difference Schemes

Methodology and Applications

Nonstandard Finite Difference Schemes
Methodology and Applications

Ronald E Mickens
Clark Atlanta University, USA

World Scientific

NEW JERSEY · LONDON · SINGAPORE · BEIJING · SHANGHAI · HONG KONG · TAIPEI · CHENNAI · TOKYO

Published by

World Scientific Publishing Co. Pte. Ltd.

5 Toh Tuck Link, Singapore 596224

USA office: 27 Warren Street, Suite 401-402, Hackensack, NJ 07601

UK office: 57 Shelton Street, Covent Garden, London WC2H 9HE

Library of Congress Cataloging-in-Publication Data
Names: Mickens, Ronald E., 1943–   author.
Title: Nonstandard finite difference schemes : methodology and applications /
    Ronald E. Mickens, Clark Atlanta University, USA.
Description: New Jersey : World Scientific, [2021] |
    Includes bibliographical references and index.
Identifiers: LCCN 2020044299 | ISBN 9789811222535 (hardcover) |
    ISBN 9789811222542 (ebook)
Subjects: LCSH: Finite differences. | Nonstandard mathematical analysis.
Classification: LCC QA431 .M4285 2021 | DDC 515/.625--dc23
LC record available at https://lccn.loc.gov/2020044299

British Library Cataloguing-in-Publication Data
A catalogue record for this book is available from the British Library.

For any available supplementary material, please visit
https://www.worldscientific.com/worldscibooks/10.1142/11891#t=suppl

Printed in Singapore

This book is dedicated
to my wife
Maria,
my son
James Williamson,
my daughter
Leah Maria.

Preface

This book is an expanded second edition of my previous volume: R. E. Mickens, *Nonstandard Finite Difference Models of Differential Equations* (World Scientific, Singapore, 1994). The purpose of the original book was to introduce a new methodology for constructing finite difference schemes for numerically integrating differential equations and this set of procedures became known as nonstandard finite difference (NSFD) schemes. However, since 1994 great progress has been made in both the understanding and the types of differential equations for which the NSFD methodology has been applied. Applications have been extended to delayed, fractional derivative, and stochastic differential equations. Further, two major summaries and reviews have been written by Professor Kailash C. Patidar. This current book is a record of the progress made on the NSFD methodology since 1994.

We estimate that more than one thousand publications have appeared which make use of the general NSFD methodology. Of interest is the fact that concepts such as dynamic consistency, denominator functions, and nonlocal discretizations of functions of the dependent variables are quickly making their way into the vocabulary of the numerical integration of differential equations often without explicit acknowledgement of their genesis within the original context of the NSFD methodology. This is a clear indication of the increasing acceptance of this set of techniques.

It should be clearly and strongly understood that the NSFD methodology is, in particular, a general philosophy as to how finite difference discretizations should be formulated and constructed for differential equations. It, in general, is not a fixed set of rules or procedures to carry out such constructions. In particular, it must be understood that each specific differential equation must be treated differently and just how this should be accomplished depends critically on the background knowledge base of the researcher and their insights into the derivation of the differential equations and its proposed applications.

This second edition reformulates just what is the NSFD methodology. Thus, the new interpretation places the principle of dynamic consistency (DC) as the essential foundation on which all the other critical features are to be based and/or derived, i.e., the calculation of denominator functions for the discrete derivatives and the nonlocal discretizations for functions of the dependent variables.

Chapters 1–8, in the original volume remain the same except for corrections and very minor additions. These Chapters still provide an excellent introduction to the genesis, construction, and history of NSFD schemes.

A new Chapter 0 has been added. It gives a brief overview of what this volume is about and gives a summary of the new materials. Chapter 9 gives a fuller and deeper explanation of the NSFD methodology. The main purpose of Chapter 10 is to summarize a large number of both ordinary and partial differential equations for which "exact finite difference schemes" are known. These results are of value, since one of the techniques used in the construction of NSFD schemes is to incorporate exact schemes of subequations into the full discretization.

Finally, Chapter 11 gives a representative sampling of the use of the NSFD methodology to construct schemes for a number of important applications in the natural and engineering sciences.

For those who wish a quick historical based introduction to the NSFD methodology and the important advances made in its techniques and applications, we suggest reading the following four articles:

1) R. E. Mickens, "Exact solutions to a finite-difference model of a nonlinear reaction-advection equation: Implications for numerical analysis," *Numerical Methods for Partial Differential Equations* **5** (1989), 313–325.

2) R. E. Mickens, "Dynamic consistency: A fundamental principle for constructing NSFD schemes for differential equations," *Journal of Difference Equations and Applications* **11** (2005), 645–653.

3) R. E. Mickens, "Numerical integration of population models satisfying conservation laws: NSFD methods," *Journal of Biological Dynamics* **1** (2007), 427–436.

4) R. E. Mickens, "Calculation of denominator functions for NSFD schemes for differential equations satisfying a positivity condition," *Numerical Methods for Partial Differential Equations* **23** (2007), 672–691.

I wish to thank the hundreds of individuals who have used, applied, and extended the NSFD methodology. In particular, I am thankful for the research productions of the following colleagues and their collaborators: Ron Buckmire, Bento M. Chen-Charpentier, James B. Cole, Matthias Ehrhardt, Saber N. Elaydi, Abba B. Gumel, Pedro M. Jordan, Hristo V. Kojouharov, Jean M.-S. Lubuma, Jorge E. Macias-Diaz, Kale Oyedeji, Kailash C. Patidar, Lih-Ing W. Roeger, and Talitha M. Washington.

Also, special thanks are extended to Imani Beverly and Bryan Brione for their help in obtaining manuscripts, books, and other technical items which greatly aided in my writing the original manuscript of this book.

As always, I am particularly grateful to Annette Rohrs for handling the typing and preliminary editorial processing of the full manuscript. Without her assistance, this volume would not exist.

Ronald E. Mickens
Atlanta, Georgia
September 2020

Contents

Chapter 0

A Second Edition ... Why?

0.1 Purpose

Our earlier book (R. E. Mickens, *Nonstandard Finite Difference Models of Differential Equations*, 1994), published a quarter century ago, had achieved great success. It introduced hundreds of researchers and practitioners to the concepts, rules, and applications of what is now recognized and called the nonstandard finite difference (NSFD) methodology which can be used to form discretizations of differential equations. We estimate that more than one thousand peer-reviewed articles and abstracts have been published on NSFD schemes and that similar numbers of presentations have been given at professional conferences, workshops, and seminars. Likewise, a number of edited volumes have been published on the application of the NSFD methodology to a broad range of differential equations modeling important phenomena in the natural and engineering sciences; see the Bibliography at the end of this book.

One purpose of this revised and expanded volume is to incorporate the great progress that has been made in both the understanding and application of the NSFD methodology since 1994. In particular, the introduction and use of the concept of "dynamical consistency" [1] often permits the direct calculation of the *a priori* unknown "denominator functions" and places severe restrictions on the structures for the nonlocal representations of functions of the dependent variables.

A second goal is to clear up and clarify certain misunderstandings and false interpretations of exactly what is the NSFD methodology and how it should be used to construct "valid" NSFD schemes. These difficulties follow in part from the general lack of full comprehension, by some, as to both the genesis and valid application of this methodology. Examples of

1

such confusions include the three cases:

1) NSFD schemes require the *a priori* knowledge of the exact solution for the differential equations of interest.

This is clearly nonsensical. If the exact solution is already known, then there are no reasons to formulate discretizations of the differential equations to calculate numerical solutions.

2) If for an ordinary differential equation, there appears a term, such as $y^2(t)$, then its discretization must take one of the forms

$$y^2(t) \rightarrow \begin{cases} y_{k+1}y_k, \\ \frac{y_{k+1}^2 + y_{k+1}y_k + y_k^2}{3}, \\ \left(\frac{y_{k+1}+y_k+y_{k-1}}{3}\right) y_k. \end{cases} \qquad (0.1.1)$$

This may or may not be correct. The particular discretizations may differ from any of these expressions.

3) The "denominator functions" [2] is (almost)

$$\phi(h) = \begin{cases} 1 - e^{-h}, \\ e^h - 1. \end{cases} \qquad (0.1.2)$$

The calculations, discussions, and figures in this revised volume will hopefully demonstrate the errors arising from taking these (and other) simplistic views of the NSFD methodology and its applications.

In the next section, we briefly explore some of the ambiguities inherent in the process of selecting a discretization for a particular differential equations. Finally, in Section 0.3, we provide comments relevant to the creation of the NSFD methodology.

0.2   Ambiguities with the Discretization Process

In this section, three rather elementary ordinary differential equations will be examined from the perspective of constructing possible discretizations.

First, consider the decay equation

$$\frac{dy(t)}{dt} = -\lambda y(t), \quad y(0) = y_0 > 0, \quad \lambda > 0. \qquad (0.2.1)$$

Using the fact that

$$y = \frac{y^2}{y}, \tag{0.2.2}$$

and a forward-Euler discretization of dy/dt, we can write

$$\frac{y_{k+1} - y_k}{h} = -\left(\frac{2\lambda y_k y_{k+1}}{y_{k+1} + y_k}\right). \tag{0.2.3}$$

Thus, solving for y_{k+1} gives the quadratic algebraic equation

$$(y_{k+1})^1 + [(2\lambda h)y_k]y_{k+1} - (y_k)^2 = 0, \tag{0.2.4}$$

which upon inspection shows that among the two solutions, one is non-negative, i.e.,

$$y_{k+1} = -(h\lambda)y_k + y_k\sqrt{1 + \lambda^2 h^2}$$
$$= \left[\sqrt{1 + \lambda^2 h^2} - h\lambda\right] y_k. \tag{0.2.5}$$

We now show that

$$0 < \sqrt{1 + \lambda^2 h^2} - h\lambda = 1. \tag{0.2.6}$$

Proof: Since for $h > 0$ and $\lambda > 0$,

$$1 + h^2\lambda^2 > h^2\lambda^2, \tag{0.2.7}$$

it follows that

$$\sqrt{1 + h^2\lambda^2} > h\lambda, \tag{0.2.8}$$

and

$$\sqrt{1 + h^2\lambda^2} - h\lambda > 0. \tag{0.2.9}$$

Also,

$$1 + 2\lambda h + \lambda^2 h^2 = (1 + \lambda h)^2 \tag{0.2.10}$$

and

$$1 + \lambda^2 h^2 < (1 + \lambda h)^2, \tag{0.2.11}$$

which implies

$$\sqrt{1 + \lambda^2 h^2} - \lambda h < 1. \tag{0.2.12}$$

From Eqs. (0.2.9) and (0.2.12), the bounds in Eq. (0.2.6) are obtained.

Inspection of Eq. (0.2.1) indicated that all its solutions are non-negative and decrease monotonous to zero [3]. The discretization, given in Eq. (0.2.5), has exactly the same feature [4].

It must be emphasized that the finite difference discretization of Eq. (0.2.1), as given by Eq. (0.2.5) is certainly not one following from using the standard procedures [5].

Now consider the differential equation

$$\frac{dy(t)}{dt} = \frac{1}{y(t)}, \quad y(0) = y_0 > 0. \tag{0.2.13}$$

Inspection of Eq. (0.2.13) indicates that its solutions are non-negative and monotonic increasing. In fact, this differential equation can be easily integrated to give

$$y(t) = \sqrt{y_0^2 + 2t}. \tag{0.2.14}$$

Now, a possible discretization is

$$\frac{y_{k+1} - y_k}{h} = \frac{2}{y_{k+1} + y_k}, \tag{0.2.15}$$

which can be solved for y_{k+1} to give

$$y_{k+1} = \sqrt{y_k^2 + 2h}. \tag{0.2.16}$$

This 1-dim mapping [4] gives solutions consistent with those of Eq. (0.2.14).

Note that the simple, standard discretization

$$\frac{y_{k+1} - y_k}{h} = \frac{1}{y_k}, \tag{0.2.17}$$

also gives solutions, y_k, in qualitative agreement with $y(t)$, the exact solution to Eq. (0.2.13).

An interesting differential equation is

$$\frac{dy(t)}{dt} = -\lambda_1 y(t)^{1/3}, \quad y(0) = y_0 > 0, \quad \lambda_1 > 0. \tag{0.2.18}$$

This ordinary differential equation has all non-negative solutions decreasing monotonous to zero. (This can be shown by using methods from the qualitative theory of differential equations [3].) Starting with

$$y^{1/3} = \frac{y}{y^{2/3}}, \tag{0.2.19}$$

a possible discretization is

$$\frac{y_{k+1} - y_k}{h} = -\lambda_1 \frac{y_{k+1}}{(y_k)^{2/3}}. \tag{0.2.20}$$

Solving for y_{k+1} gives

$$y_{k+1} = \left[\frac{y_k^{2/3}}{\lambda_1 h + y_k^{2/3}} \right] y_k. \tag{0.2.21}$$

Given $y_0 > 0$, then all solutions monotonous decrease to zero and are non-negative.

An elementary, standard discretization, such as

$$\frac{y_{k+1} - y_k}{h} = -\lambda_1 y_k^{2/3}, \quad y_0 > 0, \tag{0.2.22}$$

has certain difficulties. For example, for fixed $\lambda_1 > 0$ and $h > 0$, y_0 can be selected such that y_k is negative.

Let us now turn to the consideration of discrete representations of the first-derivative, $dy(t)/dt$. From the calculus, we find the relation [6]

$$\frac{dy(t)}{dt} = \lim_{h \to 0} \frac{y(t+h) - y(t)}{h} \tag{0.2.23}$$

and this gives rise to the so-called forward-Euler approximation

$$\frac{dy(t)}{dt} = \frac{y(t+h) - y(t)}{h} + O(h). \tag{0.2.24}$$

However, a more general expression can be constructed; one possibility is

$$D(y, t, p, h) \equiv \frac{y[(t + \psi_1(h, p)] - \psi_2(h, p) y(t)}{\phi(h, p)}, \tag{0.2.25}$$

where $p : (p_1, p_2, \ldots, p_m)$ represents m-parameters which may occur in the differential equations to be discretized. Observe that

$$\lim_{h \to 0} D(y, t, p, h) = \frac{dy(t)}{dt}, \tag{0.2.26}$$

provided

$$\psi_1(h, p) = h + O(h^2), \tag{0.2.27a}$$
$$\psi_2(h, p) = 1 + O(h^2), \tag{0.2.27b}$$
$$\phi_2(h, p) = h + O(h^2). \tag{0.2.27c}$$

For $h > 0$, $D(y, t, p, h)$ is a discrete model of the derivative $dy(t)/dt$. Therefore, for $h > 0$, the relationship

$$\frac{dy(t)}{dt} = D(y, t, p, h) + O(h) \tag{0.2.28}$$

is essentially meaningless from a computational point of view. This follows from the observation that depending on the functions $\psi_1(h,p)$, $\psi_2(h,p)$ and $\phi(h,p)$, the magnitude of $D(y,t,p,h)$ may differ greatly from the magnitude of $dy(t)/dt$. This result indicates that the task of discretizing differential equations, is in general much more complex than the application of standard procedures would lead us to believe.

Likewise, the discretizations of the term y^2 in the differential equation

$$\frac{dy(t)}{dt} = -y(t)^2, \tag{0.2.29}$$

is also not as clear cut as one would want. *A priori*, all of the following possibilities and linear combinations are potential discretizations

$$y^2 \to y_{k+1}^2, y_k^2, y_{k+1}y_k. \tag{0.2.30}$$

Perspective researchers should be attentive and perceptive to a broad range of concepts and issues relative to their research problems. They must be alert to the possibility that potential connections and relations may exist between their current work and other far flung areas of mathematics and science, and they must be able to recognize when such connections do present themselves.

For example, consider the function

$$f(x) = 1 - ax, \tag{0.2.31}$$

where a is constant. If we have the restriction

$$|ax| \ll 1, \tag{0.2.32}$$

then $f(x)$ can be approximated by either of the following forms

$$f(x) = \begin{cases} e^{-ax}, \\ \frac{1}{1+ax}. \end{cases} \tag{0.2.33}$$

An application of this result immediately presents itself. A forward-Euler finite difference scheme for Eq. (0.2.1), i.e., $dy/dt = -\lambda y$, is

$$\frac{y_{k+1} - y_k}{h} = -\lambda y_k, \tag{0.2.34a}$$

or

$$y_{k+1} = (1 - \lambda h)y_k. \tag{0.2.34b}$$

Using the second expression in Eq. (0.2.33), we can write Eq. (0.2.34b) as

$$y_{k+1} = e^{-\lambda h} y_k, \quad |\lambda h| \ll 1. \tag{0.2.35}$$

But, after several algebraic manipulations, we obtain

$$\frac{y_{k+1} - y_k}{\phi(h, \lambda)} = -\lambda y_k, \tag{0.2.36a}$$

$$\phi(h, \lambda) - \frac{1 - e^{-\lambda h}}{\lambda}. \tag{0.2.36b}$$

Note that

$$\phi(h, \lambda) = h + O(\lambda h^2), \tag{0.2.37}$$

and from the perspective of curiosity, why not examine this discretization, given by Eqs. (0.2.36) as being potentially valid for all $h > 0$?

0.3 The Nonstandard Finite Difference Methodology

A model of a physical system corresponds to the construction of an approximate mathematical representation of the system, incorporating certain important and essential features of the system, while ignoring everything else [7]. Thus, different classes of models are distinguished by which features are reproduced by the model. With this in mind, it follows that another aspect of the study of the discretization process is to obtain additional insights into the system and the consequences of its various mathematical models.

A close examination of the results presented in Section 0.2 strongly indicates that the standard finite difference discretizations, which are usually based on mathematical considerations [5], are just one of many possibilities. The task of this revised edition of my 1994 book is therefore to explicitly demonstrate the existence of another procedure for constructing finite difference schemes of differential equations. This we call the nonstandard finite difference (NSFD) methodology.

From the previous book, we retain the original Chapters 1–8, essentially unchanged, but with known typos corrected. Our goal in doing this is to provide a historical evolution and/or genesis of my work which led to a new philosophy and set of techniques as to how the various terms in differential equations should be discretized.

Users of this volume, who are new to the subject should start with Chapter 1 and work their way to the end of the book. For those with some experience with the NSFD methodology, it will be appropriate to consider only Chapters 9 and 10.

References

1. R. E. Mickens, *Journal of Difference Equations and Applications* **11**, 656–653 (2005). Dynamic consistency: A fundamental principle for constructing nonstandard finite difference schemes for differential equations.
2. R. E. Mickens, *Numerical Methods for Partial Differential Equations* **23**, 672–691 (2007). Calculation of denominator functions for NSFD schemes for differential equations satisfying a positivity condition.
3. R. E. Mickens, *Mathematical Methods for the Natural and Engineering Sciences*, 2nd Edition (World Scientific, Singapore, 2017). See Chapter 4.
4. R. E. Mickens, *Difference Equations: Theory, Applications and Advanced Topics*, 3rd Edition (CRC Press, New York, 2015).
5. M. K. Jain, *Numerical Solution of Differential Equations* (Halsted/Wiley, New York, 1984).
6. R. A. Silverman, *Essential Calculus with Applications* (Dover, New York, 1977.
7. M. B. Allen, III, I. Herrera, and G. F. Pinder, *Numerical Modeling in Science and Engineering* (Wiley, New York, 1988).

Chapter 1

Introduction

1.1 Numerical Integration

In general, a given linear or nonlinear differential equation does not have a complete solution that can be expressed in terms of a finite number of elementary functions [1–4]. A first attack on this situation is to seek approximate analytic solutions by means of various perturbation methods [5–7]. However, such procedures only hold for limited ranges of the (dimensionless) system parameters and/or the independent variables. For arbitrary values of the system parameters, at the present time, only numerical integration techniques can provide accurate numerical solutions to the original differential equations of interest. A major difficulty with numerical techniques is that a separate calculation must be formulated for each particular set of initial and/or boundary values. Consequently, obtaining a global picture of the general solution to the differential equations often requires a great deal of computation and time. However, for many problems currently being investigated in science and technology, there exist no alternatives to numerical integration.

The process of numerical integration is the replacement of a set of differential equations, both of whose independent and dependent variables are continuous, by a model for which these variables may be discrete. In general, in the model the independent variables have a one-to-one correspondence with the integers, while the dependent variables can take real values. Our major concern in this book will be the use of a particular technique for constructing discrete models of differential equations, namely, the use of finite difference methods [8–13]. No other procedures will be considered.

An important fact often overlooked in the formulation of discrete models of differential equations is that numerical integration methods should always be constructed with the help of the knowledge gained from the study of special solutions of the differential equations. For example, if the differential equations have a constant solution with a particular stability property, the discrete model should also have this constant solution with exactly the same stability property [12, 13, 14]. We will consider this issue in considerable detail in Chapters 2 and 3.

1.2 Standard Finite Difference Modeling Rules

To illustrate the construction of discrete finite difference models of differential equations, we begin with the scalar ordinary equation

$$\frac{dy}{dt} = f(y), \tag{1.2.1}$$

where $f(y)$ is, in general, a nonlinear function of y. For a uniform lattice, with step-size, $\Delta t = h$, we replace the independent variable t by

$$t \to t_k = hk, \tag{1.2.2}$$

where h is an integer, i.e.,

$$t \in \{\ldots, -2, -1, 0, 1, 2, 3, \ldots\}. \tag{1.2.3}$$

The dependent variable $y(t)$ is replaced by

$$y(t) \to y_k, \tag{1.2.4}$$

where y_k is the approximation of $y(t_k)$. Likewise, the function $f(y)$ is replaced by

$$f(y) \to f_k, \tag{1.2.5}$$

where f_k is the approximation to $f[y(t_k)]$. The simplest possibility for f_k is

$$f_k = f(y_k). \tag{1.2.6}$$

For the first derivative, any one of the following forms is suitable

$$\frac{dy}{dt} \to \begin{cases} \frac{y_{k+1} - y_k}{h}, \\ \frac{y_k - y_{k-1}}{h}, \\ \frac{y_{k+1} - y_{k-1}}{2h}. \end{cases} \tag{1.2.7}$$

These representations of the discrete first derivative are known, respectively, as the forward-Euler, backward-Euler, and central difference schemes. They follow directly from the conventional definition of the first derivative as given in a standard first course in calculus [15], i.e.,

$$\frac{dy}{dt} = \lim_{h \to 0} \begin{cases} \frac{y(t+h)-y(t)}{h}, \\ \frac{y(t)-y(t-h)}{h}, \\ \frac{y(t+h)-y(t-h)}{2h}. \end{cases} \tag{1.2.8}$$

Given a first order scalar ordinary differential equation, a discrete finite difference model is constructed by replacing in Eq. (1.2.1) the corresponding discrete terms of Eqs. (1.2.2) to (1.2.7). Thus, a simple finite difference model for Eq. (1.2.1) is given by the expression

$$\frac{y_{k+1} - y_k}{h} = f(y_k). \tag{1.2.9}$$

Other, at this stage of our discussion, equally valid discrete models are

$$\frac{y_k - y_{k-1}}{h} = f(y_k), \tag{1.2.10}$$

$$\frac{y_{k+1} - y_{k-1}}{2h} = f(y_k), \tag{1.2.11}$$

$$\frac{y_{k+1} - y_{k-1}}{2h} = f\left(\frac{y_{k+1} + y_{k-1}}{2}\right). \tag{1.2.12}$$

The model of Eq. (1.2.10) is the backward-Euler scheme. It is called an implicit scheme since for general nonlinear $f(y)$, y_k must be solved for at each value of k in terms of the previous y_{k-1}. The Eq. (1.2.11) gives the corresponding central difference scheme, while Eq. (1.2.12) is a mixed implicit, central difference scheme.

Note that all of the discrete representations reduce to the original differential equation in the appropriate limit [8].

$$h \to 0, \quad k \to \infty, \quad t_k = t = \text{fixed}. \tag{1.2.13}$$

These results indicate that the discrete modeling process has a great deal of nonuniqueness built into it.

For completeness, we give the standard discrete representation for the second derivative; it is [8]

$$\frac{d^2 y}{dt^2} \to \frac{y_{k+1} - 2y_k + y_{k-1}}{h^2}. \tag{1.2.14}$$

Again, it follows directly from the standard calculus definition of the second derivative [15].

In the next section, we will use these standard finite difference modeling rules to construct discrete representations for several rather simple, but, important in many applications, ordinary and partial differential equations.

1.3   Examples

All of the differential equations to be modeled in this section have been put in dimensional form. This means that all non-essential constants and parameters that arise in the original differential equations have been eliminated. We show how this can be done by considering two of these equations: the decay and logistic equations. The general procedure is detailed in Mickens [6].

The decay equation is

$$\frac{dx}{dt} = -\lambda x, \quad x(0) = x_0 = \text{given}, \tag{1.3.1}$$

where λ is a positive constant. Let $\bar{t}$ and y be defined as

$$\bar{t} = \lambda t, \quad y = \frac{x}{x_0}. \tag{1.3.2}$$

Substitution of these results into Eq. (1.3.1) gives the dimensionless equation

$$\frac{dy}{d\bar{t}} = -y, \quad y(0) = 1. \tag{1.3.3}$$

The so-called logistic differential equation is

$$\frac{dx}{dt} = \lambda_1 x - \lambda_2 x^2, \quad x(0) = x_0, \tag{1.3.4}$$

where λ_1 and λ_2 are positive constants. This equation can be rewritten to the form

$$\frac{dx}{\lambda_1 dt} = x \left[1 - \left(\frac{\lambda_2}{\lambda_1} \right) x \right]. \tag{1.3.5}$$

Now let

$$\bar{t} = \lambda_1 t, \quad y = \left(\frac{\lambda_2}{\lambda_1} \right) x. \tag{1.3.6}$$

Substitution of Eq. (1.3.6) in to Eq. (1.3.5) gives the following dimensionless equation

$$\frac{dy}{d\bar{t}} = y(1 - y), \quad y(0) = \left(\frac{\lambda_2}{\lambda_1} \right) x_0. \tag{1.3.7}$$

Observe that in dimensionless form, both the decay and logistic equations have no arbitrary parameters.

Independently, as to whether we wish to numerically integrate a differential equation or not, it should always be transformed to a dimensionless form. Note that the "physical" original differential equation connects the derivatives of a physical variable such as distance or current and its relations to the various physical parameters, while the dimensionless transformed equation relates the various derivatives of a "mathematical" variable and associated constants that appear in the equation.

1.3.1 *Decay Equation*

The decay differential equation is

$$\frac{dy}{dt} = -y. \tag{1.3.8}$$

The direct forward-Euler scheme is

$$\frac{y_{k+1} - y_k}{h} = -y_k. \tag{1.3.9}$$

A discrete model can also be constructed by using a symmetric expression for the linear term y in Eq. (1.3.8). For example

$$\frac{y_{k+1} - y_k}{h} = -\left(\frac{y_{k+1} + y_k}{2}\right). \tag{1.3.10}$$

The use of the central difference for the first derivative gives

$$\frac{y_{k+1} - y_{k-1}}{2h} = y_k. \tag{1.3.11}$$

Another central difference scheme is

$$\frac{y_{k+1} - y_{k-1}}{2h} = -\left(\frac{y_{k+1} + y_k + y_{k-1}}{3}\right). \tag{1.3.12}$$

Likewise, a backward-Euler model is

$$\frac{y_k - y_{k-1}}{h} = -y_k, \tag{1.3.13}$$

which can be written as

$$\frac{y_{k+1} - y_k}{h} = -y_{k+1}. \tag{1.3.14}$$

It is clear that these discrete models are all different. For example, writing them in reduced form gives the following results for the indicated equations:

Eq. (1.3.9):

$$y_{k+1} = (1 - h)y_k, \tag{1.3.15}$$

Eq. (1.3.10):

$$y_{k+1} = \frac{(1 - h/2)}{(1 + h/2)} y_k, \tag{1.3.16}$$

Eq. (1.3.11):

$$y_{k+2} + 2hy_{k+1} - y_k = 0, \tag{1.3.17}$$

Eq. (1.3.12):

$$\left(1 + \frac{2h}{3}\right) y_{k+2} + \left(\frac{2h}{3}\right) y_{k+1} - \left(1 - \frac{2h}{3}\right) y_k = 0, \tag{1.3.18}$$

Eq. (1.3.14):

$$y_{k+1} = \left(\frac{1}{1+h}\right) y_k. \tag{1.3.19}$$

Note that Eqs. 1.3.15, (1.3.16) and (1.3.19) are first-order linear difference equations, while Eqs. (1.3.17) and (1.3.18) are second-order linear difference equations. Further, observe that all the equations of a given order have different constant coefficients for fixed step-size h. This implies that Eqs. (1.3.16) to (1.3.19) have different solutions [16]. Consequently, we must conclude that each of the above discrete models of the decay equation gives unique numerical solutions that differ from that of the other discrete models.

Again observe that each of these discrete models has coefficients that depend on the step-size h. This leaves open the possibility that the solution behaviors may vary with h. In the next chapter, we will see that this is in fact the situation.

1.3.2   *Logistic Equation*

The forward-Euler scheme for the logistic differential equation

$$\frac{dy}{dt} = y(1-y), \tag{1.3.20}$$

is

$$\frac{y_{k+1} - y_k}{h} = y_k(1 - y_k). \tag{1.3.21}$$

The corresponding backward-Euler and central difference schemes are, respectively,

$$\frac{y_{k+1} - y_k}{h} = y_{k+1}(1 - y_{k+1}), \tag{1.3.22}$$

$$\frac{y_{k+1} - y_k}{2h} = y_k(1 - y_k). \tag{1.3.23}$$

The above three equations can be rewritten to the following forms:
Eq. (1.3.21):

$$y_{k+1} = (1+h)y_k - h(y_k)^2, \tag{1.3.24}$$

Eq. (1.3.22):

$$h(y_{k+1})^2 + (1-h)y_{k+1} - y_k = 0, \tag{1.3.25}$$

Eq. (1.3.23):

$$y_{k+2} = y_k + 2hy_{k+1}(1 - y_{k+1}). \tag{1.3.26}$$

Examination of these three equations shows that while all of them are nonlinear difference equations, the forward- and backward-Euler schemes are first-order, while the central scheme is second-order. The forward-Euler and the central schemes are explicit, in the sense that the value of y_k can be determined from its, respective, values at y_{k-1} and, y_{k-1} and y_{k-2}. However, the backward-Euler scheme requires the solution of a quadratic equation at each step. The existence and uniqueness theorems for difference equations [16] lead to the conclusion that these three finite difference schemes for the logistic equation have different solutions and the nature of the solutions may change as a function of the step-size h.

1.3.3 *Harmonic Oscillator*

The dimensionless, damped harmonic oscillator equation is

$$\frac{d^2y}{dt^2} + 2e\frac{dy}{dt} + y = 0. \tag{1.3.27}$$

Consider first the case for which $e = 0$, i.e., no damping is present. For this situation the equation of motion is

$$\frac{d^2y}{dt^2} + y = 0. \tag{1.3.28}$$

The simplest discrete model is one that uses a central difference for the discrete second derivative; this scheme is

$$\frac{y_{k+1} - 2y_k + y_{k-1}}{h^2} + y_k = 0. \tag{1.3.29}$$

Two other models that use a symmetric form for the linear term y in the differential equation are

$$\frac{y_{k+1} - 2y_k + y_{k-1}}{h^2} + \frac{y_{k+1} + y_{k-1}}{2} = 0, \tag{1.3.30}$$

and

$$\frac{y_{k+1} - 2y_k + y_{k-1}}{h^2} + \frac{y_{k+1} + y_k + y_{k-1}}{3} = 0. \tag{1.3.31}$$

Two discrete models having nonsymmetric forms for the linear term y are the following

$$\frac{y_{k+1} - 2y_k + y_{k-1}}{h^2} + y_{k-1} = 0, \tag{1.3.32}$$

$$\frac{y_{k+1} - 2y_k + y_{k-1}}{h^2} + y_{k+1} = 0. \tag{1.3.33}$$

All four models are linear, second-order difference equations with constant (for h = fixed) coefficients. These coefficients differ from one model to the next. Consequently, we again must conclude that each discrete model will provide a different numerical solution.

The same conclusion is reached when the damping term in Eq. (1.3.27) is present, i.e., $e > 0$. For example, the following three equations correspond to using a centered discrete second-order derivative, a centered linear term, and, respectively, forward-Euler, backward-Euler and centered representations for the discrete first-order derivative:

$$\frac{y_{k+1} - 2y_k + y_{k-1}}{h^2} + 2e\left(\frac{y_{k+1} - y_k}{h}\right) + y_k = 0, \qquad (1.3.34)$$

$$\frac{y_{k+1} - 2y_k + y_{k-1}}{h^2} + 2e\left(\frac{y_k - y_{k-1}}{h}\right) + y_k = 0, \qquad (1.3.35)$$

$$\frac{y_{k+1} - 2y_k + y_{k-1}}{h^2} + 2e\left(\frac{y_{k+1} - y_{k-1}}{2h}\right) + y_k = 0. \qquad (1.3.36)$$

All of these models are second-order and linear; however, they clearly have different constant coefficients which implies that they have different solutions.

1.3.4   *Unidirectional Wave Equation*

A linear equation that describes waves propagating along the z-axis with unit velocity is the unidirectional wave equation

$$u_t + u_x = 0, \qquad (1.3.37)$$

where $u = u(x,t)$ and

$$u_t \equiv \frac{\partial u}{\partial t}, \quad u_x \equiv \frac{\partial u}{\partial x}. \qquad (1.3.38)$$

Denote the discrete space and time variables by

$$t_k = (\Delta t)k, \quad x_m = (\Delta x)m, \qquad (1.3.39)$$

where

$$k \in \{\ldots, -2, -1, 0, 1, 2, 3, \ldots\}, \qquad (1.3.40)$$

$$m \in \{\ldots, -2, -1, 0, 1, 2, 3, \ldots\}. \qquad (1.3.41)$$

Thus, the discrete approximation to $u(x,t)$ is

$$u(x,t) \rightarrow u_m^k, \qquad (1.3.42)$$

and the corresponding discrete first-derivatives are [8, 12]

$$\frac{\partial u}{\partial t} \rightarrow \begin{cases} \frac{u_m^{k+1} - u_m^k}{\Delta t}, \\ \frac{u_m^k - u_m^{k-1}}{\Delta t}, \\ \frac{u_m^{k+1} - u_m^{k-1}}{2\Delta t}, \end{cases} \tag{1.3.43}$$

and

$$\frac{\partial u}{\partial x} \rightarrow \begin{cases} \frac{u_{m+1}^k - u_m^k}{\Delta x}, \\ \frac{u_m^k - u_{m-1}^k}{\Delta x}, \\ \frac{u_{m+1}^k - u_{m-1}^k}{2\Delta x}. \end{cases} \tag{1.3.44}$$

Various discrete models can be constructed by selecting a particular representation for the discrete time-derivative and a second particular representation for the discrete space-derivative. The following four cases illustrate this procedure.

(i) Forward-Euler time-derivative and forward-Euler space-derivative:

$$\frac{u_m^{k+1} - u_m^k}{\Delta x} + \frac{u_{m+1}^k - u_m^k}{\Delta x} = 0; \tag{1.3.45}$$

(ii) forward-Euler time-derivative and backward-Euler space-derivative:

$$\frac{u_m^{k+1} - u_m^k}{\Delta t} + \frac{u_m^k - u_{m-1}^k}{\Delta t} = 0; \tag{1.3.46}$$

(iii) forward-Euler time-derivative and central difference space-derivative:

$$\frac{u_m^{k+1} - u_m^k}{\Delta t} + \frac{u_{m+1}^k - u_{m-1}^k}{2\Delta x} = 0; \tag{1.3.47}$$

(iv) an implicit scheme with forward-Euler for the time-derivative and backward-Euler for the space-derivative:

$$\frac{u_m^{k+1} - u_m^k}{\Delta t} + \frac{u_m^{k+1} - u_{m-1}^{k+1}}{\Delta x} = 0. \tag{1.3.48}$$

Clearly, a number of other discrete models can be easily constructed.

All of the above equations are linear partial difference equations with constant coefficients (for fixed Δt and Δx). The models of Eqs. (1.3.45), (1.3.46) and (1.3.48) are of first-order in both the discrete time and space variables. Equation (1.3.47) is first-order in the discrete time variable, but, is of second-order in the discrete space variable. Again, by inspection all four models are different and thus will give different numerical solutions to the original partial differential equation.

1.3.5 *Diffusion Equation*

The simple linear diffusion partial differential equation, in dimensionless form, is

$$u_t = u_{xx}, \quad u = u(x, t). \tag{1.3.49}$$

The standard explicit form for this equation is given by the expression

$$\frac{u_m^{k+1} - u_m^k}{\Delta t} = \frac{u_{m+1}^k - 2u_m^k + u_{m-1}^k}{(\Delta x)^2}, \tag{1.3.50}$$

while the standard implicit form is

$$\frac{u_m^{k+1} - u_m^k}{\Delta t} = \frac{u_{m+1}^{k+1} - 2u_m^{k+1} + u_{m-1}^{k+1}}{(\Delta x)^2}. \tag{1.3.51}$$

While both of these equations are linear, partial difference equations that are first-order in the discrete time and second-order in the discrete space variables, they are not identical and consequently their solutions will give different numerical solutions to the diffusion equation.

1.3.6 *Burgers' Equation*

The inviscid Burgers' partial differential equation is [13]

$$u_t + u u_x = 0, \quad u = u(x, t). \tag{1.3.52}$$

The following four equations are examples of discrete models that can be constructed for Eq. (1.3.52) using the standard finite difference rules.

(i) Forward-Euler for the time-derivative and forward-Euler for the space-derivative:

$$\frac{u_m^{k+1} - u_m^k}{\Delta t} + u_m^k \left(\frac{u_{m+1}^k - u_m^k}{\Delta x} \right) = 0; \tag{1.3.53}$$

(ii) forward-Euler for the time-derivative and implicit forward-Euler for the space derivative:

$$\frac{u_m^{k+1} - u_m^k}{\Delta t} + u_m^k \frac{u_{m+1}^{k+1} - u_m^{k+1}}{\Delta x} = 0; \tag{1.3.54}$$

(iii) central difference schemes for both the time- and space-derivatives:

$$\frac{u_m^{k+1} - u_m^{k-1}}{2\Delta t} + u_m^k \left(\frac{u_{m+1}^k - u_{m-1}^k}{2\Delta x} \right) = 0; \tag{1.3.55}$$

(iv) forward-Euler for the time-derivative and backward-Euler for the space-derivative:

$$\frac{u_m^{k+1} - u_m^k}{\Delta x} + u_m^k \left(\frac{u_m^k - u_{m-1}^k}{\Delta x} \right) = 0. \tag{1.3.56}$$

Note that in the limits

$$k \to \infty, \quad \Delta t \to 0, \quad t_k = t = \text{fixed}, \tag{1.3.57}$$

$$m \to \infty, \quad \Delta x \to 0, \quad x_k = x = \text{fixed}, \tag{1.3.58}$$

all of these difference schemes reduce to the inviscid Burgers' equation. However, inspection shows that for finite Δt and Δx these four partial difference equations are not identical. This fact leads to the conclusion that they will give numerical solutions that differ from each other.

1.4 Critique

The major result coming from the analysis of the previous two sections is the ambiguity of the modeling process for the construction of discrete finite difference models of differential equations. The use of the standard rules does not lead to a unique discrete model. Consequently, one of the questions before us is which, if any, of the standard finite difference schemes should be used to obtain numerical solutions for a particular differential equation? Another very important issue is the relationship between the solutions to a given discrete model and that of the corresponding differential equation. As indicated in Section 1.3, this connection may be tenuous. This and related matters lead to the study of numerical instabilities which is the subject of Chapter 2.

Once a discrete model is selected, the calculation of a numerical solution requires the choice of a time and/or space step-size. How should this be done? For problems in the sciences and engineering, the value of the step-sizes must be determined such that the physical phenomena of interest can be resolved on the scale of the computational grid or lattice [12–14]. However, suppose one is interested in the long-time or asymptotic-in-space behavior of the solution; can the step-sizes be taken as large as one wishes?

The general issue can be summarized as follows: Consider the scalar differential equation of Eq. (1.2.1). Select a finite difference scheme to numerically integrate this equation. At the grid point $t = t_k$, denote by $y(t_k)$ the solution to the differential equation and by $y_k(h)$ the solution to the discrete model. (Note that the numerical solution is written in such a way as to indicate that its value depends on the step-size, h.) What is the relationship between $y(t_k)$ and $y_k(h)$? In particular, how does this relationship change as h varies?

Numerical instabilities exist in the numerical solutions whenever the qualitative properties of $y_k(h)$ differ from those of $y(t_k)$. One of the tasks of this book will be to eliminate the elementary numerical instabilities that can arise in the finite difference models of differential equations. Our general goal will be the construction of discrete models whose solutions have the same qualitative properties as that of the corresponding differential equation for all step-sizes. We have not entirely succeeded in this effort, but, progress has definitely been made.

A final comment should be made on the issue of chaos and differential equations. In the past several decades, much effort has been devoted to the study of "mathematical chaos" as it occurs in the solutions of deterministic systems modeled by coupled differential equations [17–19]. Experimentally, chaotic-like behavior has been measured in fluid phenomena [19, 20], chemical reactions [21], nonlinear electrical and mechanical oscillations [22, 23] and in biomedical systems [24]. In this book, we do not directly address these issues. Our task is to formulate discrete finite difference models of differential equations that have numerical solutions which reflect accurately the underlying mathematical structures of the solutions to the differential equations and these may possess chaotic behaviors.

References

1. S. L. Ross, *Differential Equations* (Blaisdell; Waltham, MA; 1964).
2. M. Humi and W. Miller, *Second Course in Ordinary Differential Equations for Scientists and Engineers* (Springer-Verlag, New York, 1988).
3. D. Zwillinger, *Handbook of Differential Equations* (Academic Press, Boston, 1989).
4. D. Zwillinger, *Handbook of Integration* (Jones and Bartlett, Boston, 1992).
5. C. M. Bender and S. A. Orszag, *Advanced Mathematical Methods for Scientists and Engineers* (McGraw-Hill, New York, 1978).
6. R. E. Mickens, *Nonlinear Oscillations* (Cambridge University Press, New York, 1981).
7. J. Kevorkian and J. D. Cole, *Perturbation Methods in Applied Mathematics* (Springer-Verlag, New York, 1981).
8. F. B. Hildebrand, *Finite-Difference Equations and Simulations* (Prentice-Hall; Englewood Cliffs, NJ; 1968).
9. M. K. Jain, *Numerical Solution of Differential Equations* (Halsted Press/John Wiley and Sons, New York, 2nd edition, 1984).

10. J. M. Ortega and W. G. Poole, Jr., *An Introduction to Numerical Methods for Differential Equations* (Pitman; Marshfield, MA; 1981).

11. J. C. Butcher, *The Numerical Analysis of Ordinary Differential Equations: Runge-Kutta and General Linear Methods* (Wiley-Interscience, New York, 1987).

12. D. Greenspan and V. Casulli, *Numerical Analysis for Applied Mathematics, Science and Engineering* (Addison-Wesley; Redwood City, CA; 1988).

13. M. B. Allen III, I. Herrera and G. F. Pinder, *Numerical Modeling in Science and Engineering* (Wiley-Interscience, New York, 1988).

14. D. Potter, *Computational Physics* (Wiley-Interscience, New York, 1973).

15. J. Marsden and A. Weinstein, *Calculus I* (Springer-Verlag, New York , 1984); section 1.3.

16. R. E. Mickens, *Difference Equations: Theory and Applications* (Van Nostrand Reinhold, New York, 2nd edition, 1990).

17. S. Wiggins, *Introduction to Applied Nonlinear Dynamical Systems and Chaos* (Springer-Verlag, New York, 1990).

18. R. L. Devaney, *An Introduction to Chaotic Dynamical Systems* (Benjamin/Cummings; Menlo Park, CA; 1986).

19. B.-L. Hao, editor, *Chaos* (World Scientific, Singapore, 1984).

20. G. L. Baker and J. P. Gollub, *Chaotic Dynamics* (Cambridge University Press, New York, 1990).

21. S. K. Scott, *Chemical Chaos* (Clarendon Press, Oxford, 1991).

22. F. C. Moon, *Chaotic Vibrations* (Wiley-Interscience, New York, 1987).

23. J. M. T. Thompson and H. B. Stewart, *Nonlinear Dynamics and Chaos* (Wiley, New York, 1986).

24. B. J. West, *Fractal Physiology and Chaos in Medicine* (World Scientific, Singapore, 1990).

Chapter 2

Numerical Instabilities

2.1 Introduction

A discrete model of a differential equation is said to have numerical instabilities if there exist solutions to the finite difference equations that do not
correspond qualitatively to any of the possible solutions of the differential
equation. It is doubtful if a precise definition can ever be stated for the
general concept of numerical instabilities. This is because it is always possible, in principle, for new forms of numerical instabilities to arise when new
nonlinear differential equations are discretely modeled. The concept, as we
will use it in this book, will be made clearer in the material to be discussed
in this Chapter. Numerical instabilities are an indication that the discrete
equations are not able to model the correct mathematical properties of the
solutions to the differential equations of interest. The fundamental reason
for the existence of numerical instabilities is that the discrete models of differential equations have a larger parameter space than the corresponding
differential equations. This can be easily demonstrated by the following
argument. Assume that a given dynamic system is described in terms of
the differential equation

$$\frac{dy}{dt} = f(y, \lambda), \tag{2.1.1}$$

where λ denotes the n-dimensional parameter vector that defines the system. A discrete model for Eq. (2.1.1) takes the form

$$y_{k+1} = F(y_k, \lambda, h) \tag{2.1.2}$$

where $h = \Delta t$ is the time step-size. Note that the function F contains
$(n+1)$ parameters; this is because h can now be regarded as an additional
parameter. The solutions to Eqs. (2.1.1) and (2.1.2) can be written, re-
spectively as $y(t, \lambda)$ and $y_k(\lambda, h)$. Even if $y(t, \lambda)$ and $y_k(\lambda, h)$ are "close" to

each other for a particular value of h, say $h = h_1$. If h is changed to a new value, say $h = h_2$, the possibility exists that $y_k(\lambda, h_2)$ differs greatly from $y_k(\lambda, h_1)$ both qualitatively and quantitatively. The detailed study of what actually occurs relies on the use of bifurcation theory [1, 2, 3].

The purpose of this Chapter is to consider a number of differential equations, construct several discrete models of them, and compare the properties of the solutions to the difference equations to the corresponding properties of the original differential equations. Any discrepancies found are indications of numerical instabilities. We end the Chapter with a summary and discussion of the elementary numerical instabilities.

2.2   Decay Equation

The decay differential equation is

$$\frac{dy}{dt} = -y. \tag{2.2.1}$$

Even if we did not know how to solve this equation exactly, its general solution behavior could be obtained by knowledge of the fact that for $y > 0$, the derivative is negative, while for $y < 0$, the derivative is positive. Also, $y = 0$ is a solution of the differential equation. Consequently, all solutions have the forms as shown in Figure 2.2.1(b). For the initial condition

$$y(t_0) = y_0, \tag{2.2.2}$$

the exact general solution is

$$y(t) = y_0 e^{-(t-t_0)}. \tag{2.2.3}$$

From either Figure 2.2.1(b) or Eq. (2.2.3), it can be concluded that all solutions monotonically decrease (in absolute value) to zero as $t \to \infty$.

The forward-Euler scheme for the decay equation is

$$\frac{y_{k+1} - y_k}{h} = -y_k, \tag{2.2.4}$$

where $h = \Delta t$ is the step-size. It can be rewritten to the form

$$y_{k+1} = (1 - h)y_k, \tag{2.2.5}$$

which shows it to be a first-order, linear difference equation with constant coefficients. Its solution is

$$y_k = y_0(1 - h)^k. \tag{2.2.6}$$

Note that the behavior of the solution depends on the value of $r(h) = 1 - h$ which is plotted in Figure 2.2.2. Referring to Figure 2.2.3, the following conclusions are reached:

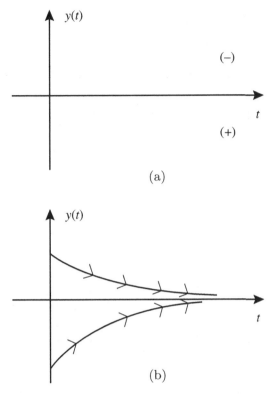

Fig. 2.2.1 The decay equation, (a) Regions where the derivative has a constant sign, (b) Typical trajectories.

(i) If $0 < h < 1$, then y decreases monotonically to zero.

(ii) If $h = 1$, then for $k \geq 1$, the solution is identically zero.

(iii) If $1 < h < 2$, y_k decreases to zero with an oscillating (change in sign) amplitude.

(iv) If $h = 2$, then y oscillates with a constant amplitude. The solution has period-two.

(v) If $h > 2$, y_k oscillates with an exponentially increasing amplitude.

Note that it is only for cases (i) and (ii) that we obtain a y_k that has the same qualitative behavior as the actual solution to the decay equation, namely, a monotonic decrease to zero. Quantitative agreement between $y(t)$ and y_k can be gotten by choosing h small, i.e., $0 < h \ll 1$.

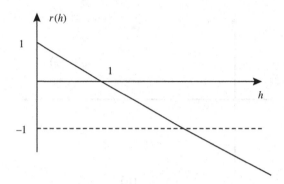

Fig. 2.2.2   Plot of $r(h) = 1 - h$.

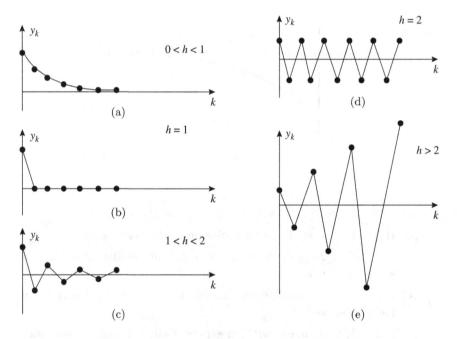

Fig. 2.2.3   Plots of solutions to $y_{k+1} = (1 - h)y_k$.

The solution behaviors exhibited, in particular, by Figures 2.2.3(c), (d) and (e), we call numerical instabilities.

We now consider a forward-Euler scheme with a symmetric form for the linear term; it is given by the expression

$$\frac{y_{k+1} - y_k}{h} = -\left(\frac{y_{k+1} + y_k}{2}\right). \tag{2.2.7}$$

Solving for y_{k+1} gives

$$y_{k+1} = \left(\frac{2-h}{2+h}\right) y_k. \tag{2.2.8}$$

Again, this is a first-order, linear difference equation with constant coefficients. Its solution behavior is dependent on the value of

$$r(h) = \frac{2-h}{2+h}, \tag{2.2.9}$$

which is plotted in Figure 2.2.4. Since

$$|r(h)| < 1, \quad 0 < h < \infty, \tag{2.2.10}$$

it follows that all solutions of Eq. (2.2.8)

$$y_k = y_0[r(h)]^k, \tag{2.2.11}$$

decrease to zero with $k \to \infty$. However, only for $0 < h < 2$ is the decrease monotonic. If $h > 2$, the solution is oscillatory with an amplitude that decreases exponentially. See Figure 2.2.5.

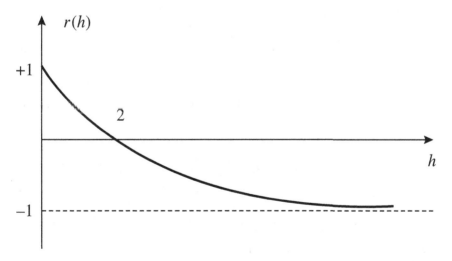

Fig. 2.2.4　Plot of $r(h) = \frac{2-h}{2+h}$.

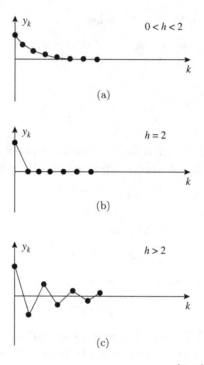

Fig. 2.2.5   Plots of solutions to $y_{k+1} = \left(\frac{2-h}{2+h}\right)y_k$.

The backward-Euler scheme for the decay equation is

$$\frac{y_k - y_{k-1}}{h} = -y_k, \qquad (2.2.12)$$

or

$$y_{k+1} = \left(\frac{1}{1+h}\right)y_k. \qquad (2.2.13)$$

Since

$$0 < \frac{1}{1+h} < 1, \quad 0 < h < \infty, \qquad (2.2.14)$$

it follows that all the solutions of Eq. (2.2.13), i.e.,

$$y_k = y_0\left(\frac{1}{1+h}\right)^k, \qquad (2.2.15)$$

decrease (in magnitude) to zero monotonically for any step-size.

The central difference scheme is

$$\frac{y_{k+1} - y_{k-1}}{2h} = -y_k. \qquad (2.2.16)$$

This is a second-order, linear difference equation

$$y_{k+2} + (2h)y_{k+1} - y_k = 0, \tag{2.2.17}$$

having constant coefficients. Its solution is

$$y_k = C_1(r_+)^k + C_2(r_-)^k, \tag{2.2.18}$$

where C_1 and C_2 are arbitrary constants, and (r_+, r_-) are solutions to the characteristic equation for Eq. (2.2.17)

$$r^2 + (2h)r - 1 = 0. \tag{2.2.19}$$

$$r_+(h) = -h + \sqrt{1 + h^2}, \tag{2.2.20a}$$

$$r_-(h) = -h - \sqrt{1 + h^2}. \tag{2.2.20b}$$

An easy set of calculations shows the following to be true:

(i) $r_-(h) < -1$, $0 < h < \infty$;
(ii) $r_-(h) = -2h + O\left(\frac{1}{2h}\right)$, for $h \to \infty$;
(iii) $0 < r_+(h) < 1$, $0 < h < \infty$;
(iv) $r_+(h) = \frac{1}{2h} + O\left(\frac{1}{h^3}\right)$, for $h \to \infty$.

These facts lead to the conclusion that the second term on the right-hand side of Eq. (2.2.18) oscillates with an amplitude that increases exponentially, while the first term decreases monotonically to zero. A typical solution trajectory is shown in Figure 2.2.6. Since this behavior holds for any step-size $h > 0$, we see that the central difference scheme has numerical instabilities regardless of the value of h.

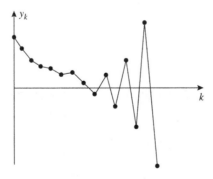

Fig. 2.2.6   Plot of a typical solution to $\frac{y_{k+1} - y_{k-1}}{2h} = -y_k$.

In summary, the four discrete models of the decay equation only give the correct qualitative behavior for the numerical solution if the following conditions are satisfied for the step-size h;

   (a) forward-Euler: $0 < h < 1$;
   (b) forward-Euler with symmetric linear term: $0 < h < 2$;
   (c) backward-Euler: $h > 0$;
   (d) central difference: no value of h.

From these results, we conclude that the central difference scheme has numerical instabilities for all step-size values; the forward-Euler schemes provide useful discrete models if limitations are placed on the step-size; and the backward-Euler scheme can be used for any (positive) step-size. Except for the central difference scheme, the other three discrete models will give excellent quantitative numerical solutions if h is made small enough, i.e., $0 < h \ll 1$ [4].

2.3   Harmonic Oscillator

The harmonic oscillator differential equation

$$\frac{d^2y}{dt^2} + y = 0, \tag{2.3.1}$$

is characterized by the fact that all its solutions are periodic [5, 6]

$$y(t) = C_1 \cos t + C_2 \sin t = Ce^{it} + C^* e^{-it}, \tag{2.3.2}$$

where C_1 and C_2 are arbitrary constants, and C is a complex-valued constant.

The straightforward central difference scheme is

$$\frac{y_{k+1} - 2y_k + y_{k-1}}{h^2} + y_k = 0, \tag{2.3.3}$$

which can be rewritten to the form

$$y_{k+1} - (2 - h^2)y_k + y_{k-1} = 0. \tag{2.3.4}$$

This is a second-order, linear difference equation whose solution is

$$y_k = D_1(r_+)^k + D_2(r_-)^k, \tag{2.3.5}$$

where D_1 and D_2 are constants, and (r_+, r_-) are solutions to the characteristic equation

$$r^2 - 2\left(1 - \frac{h^2}{2}\right)r + 1 = 0. \tag{2.3.6}$$

Solving Eq. (2.3.6) gives

$$r_+(h) = \left(1 - \frac{h^2}{2}\right) + \left(\frac{h}{2}\right)\sqrt{h^2 - 4}, \tag{2.3.7a}$$

$$r_-(h) = \left(1 - \frac{h^2}{2}\right) - \left(\frac{h}{2}\right)\sqrt{h^2 - 4}. \tag{2.3.7b}$$

For $0 < h < 2$, $r_+(h)$ and $r_-(h)$ are complex-valued with

$$r_+(h) = [r_-(h)]^* = \left(1 - \frac{h^2}{2}\right) + \left(\frac{ih}{2}\right)\sqrt{4 - h^2}. \tag{2.3.8}$$

They also have magnitude one since

$$|r_+(h)|^2 = |r_-(h)|^2 = \left(1 - \frac{h^2}{2}\right)^2 + \left(\frac{h^2}{4}\right)(4 - h)^2 = 1. \tag{2.3.9}$$

Hence, for $0 < h < 2$, $r_+(h)$ and $r_-(h)$ have the representations

$$r_+(h) = [r_-(h)]^* = e^{i\phi(h)}, \tag{2.3.10}$$

$$\tan\phi(h) = \frac{(h/2)\sqrt{4 - h^2}}{\left(1 - \frac{h^2}{2}\right)}. \tag{2.3.11}$$

Consequently, the general solution to Eq. (2.3.4), for $0 < h < 2$, is

$$y_k = D_1 e^{i\phi(h)k} + D_2 e^{-i\phi(h)h}. \tag{2.3.12}$$

If $h = 2$, then

$$r_+(2) = r_-(2) = -1, \tag{2.3.13}$$

and the general solution is

$$y_k = (D_1 + D_2 k)(-l)^k. \tag{2.3.14}$$

For $h > 2$, $r_+(h)$ and $r_-(h)$ are both real and given by the expressions

$$r_+(h) = -\left(\frac{h^2}{2} - 1\right) + \left(\frac{h}{2}\right)\sqrt{h^2 - 4}, \tag{2.3.15a}$$

$$r_-(h) = -\left(\frac{h^2}{2} - 1\right) - \left(\frac{h}{2}\right)\sqrt{h^2 - 4}. \tag{2.3.15b}$$

It follows from Eq. (2.3.15b) that

$$r_-(h) < -1, \quad h > 2, \tag{2.3.16}$$

and from the characteristic Eq. (2.3.6) that

$$r_+(h)r_-(h) = 1. \tag{2.3.17}$$

This implies that $r_+(h)$ must have a negative sign with a magnitude less than one, i.e.,

$$-1 < r_+(h) < 0, \quad h > 2. \tag{2.3.18}$$

Figure 2.3.1 gives the behavior of $r_+(h)$ and $r_-(h)$ as a function of h. Thus, for this case

$$y_k = [D_1|r_+(h)|^k + D_2|r - (h)|^k](-1)^k, \tag{2.3.19}$$

and we conclude that y_k will increase exponentially with an oscillating amplitude.

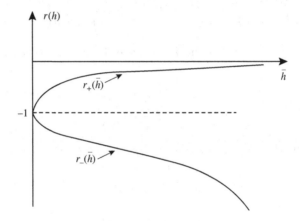

Fig. 2.3.1   Plots of $r_+(\bar{h})$ and $r_-(\bar{h})$ from Eqs. (2.3.15a) and (2.3.15b); $\bar{h} = h - 2$.

Putting all these results together, we observe that the straightforward central difference scheme has a solution with the same qualitative behavior as the harmonic oscillator differential equation only if the step-size is restricted to the interval $0 < h < 2$.

We now consider two central difference schemes for which the linear y term is modeled with a non-symmetric form. These discrete models are

$$\frac{y_{k+1} - 2y_k + y_{k-1}}{h^2} + y_{k-1} = 0, \tag{2.3.20}$$

and

$$\frac{y_{k+1} - 2y_k + y_{k-1}}{h^2} + y_{k+1} = 0. \tag{2.3.21}$$

The characteristic equation for Eq. (2.3.20) is

$$r^2 - 2r + (l + h^2) = 0, \tag{2.3.22}$$

with solutions

$$r_+(h) = [r_-(h)]^* = 1 + ih. \tag{2.3.23}$$

These can also be rewritten in the polar form

$$r_+(h) = \sqrt{1 + h^2}\, e^{i\phi(h)}, \tag{2.3.24}$$

$$\tan \phi(h) = h. \tag{2.3.25}$$

Note that the two roots are complex valued for all $h > 0$ and that they have magnitudes that are greater than one. As a consequence, all the solutions of this discrete model are oscillatory, but, they have an amplitude that increases exponentially.

Likewise, the characteristic equation for Eq. (2.3.21) is

$$r^2 - \left(\frac{2}{1+h^2}\right) r + \left(\frac{1}{1+h^2}\right) = 0. \tag{2.3.26}$$

Its solutions are again complex valued for all $h > 0$; they are

$$r_+(h) = [r_-(h)]^* = \frac{1+ih}{1+h^2} = \frac{1}{\sqrt{1+h^2}} e^{i\phi(h)}, \tag{2.3.27}$$

where $\phi(h)$ is given by the relation of Eq. (2.3.25). Therefore, we conclude that, for $h > 0$, all the solutions of Eq. (2.3.21) are oscillatory with an amplitude that decreases exponentially.

We now examine the properties of a central difference scheme having a symmetric form for the linear term y. For the first example, we consider the following discrete model

$$\frac{y_{k+1} - 2y_k + y_{k-1}}{h^2} + \frac{y_{k+1} + y_{k-1}}{2} = 0. \tag{2.3.28}$$

Its characteristic equation is

$$r^2 - \left[\frac{2}{1+\frac{h^2}{2}}\right] r + 1 = 0, \tag{2.3.29}$$

with roots

$$r_\pm(h) = \left[\frac{1}{1+\frac{h^2}{2}}\right]\left[1 \pm ih\sqrt{1+\frac{h^2}{4}}\right]. \tag{2.3.30}$$

Note that

$$r_+(h) = [r_-(h)]^*, \quad h > 0, \tag{2.3.31}$$

$$|r_+(h)| = |r_-(h)| = 1; \tag{2.3.32}$$

consequently,

$$r_+ = r_-^* = e^{i\phi(h)}, \tag{2.3.33}$$

$$\tan \phi(h) = h\sqrt{1 + \frac{h^2}{4}}. \tag{2.3.34}$$

Since

$$y_k = E(r_+)^k + E^*(r_+^*)^k, \tag{2.3.35}$$

where E is an arbitrary complex-valued constant, we conclude that all solutions to this discrete model oscillate with constant amplitude for $h > 0$.

The second example has a completely symmetric discrete expression for the linear term; it is

$$\frac{y_{k+1} - 2y_k + y_{k-1}}{h^2} + \frac{y_{k+1} + y_k + y_{k-1}}{3} = 0. \tag{2.3.36}$$

The corresponding characteristic equation is

$$r^2 - 2\left[\frac{1 - \frac{h^2}{6}}{1 + \frac{h^2}{3}}\right]r + 1 = 0, \tag{2.3.37}$$

with the roots

$$r_\pm(h) = \left[\frac{1}{1 + \frac{h^2}{3}}\right]\left\{\left(1 - \frac{h^2}{6}\right) \pm ih\sqrt{1 + \frac{h^2}{12}}\right\}. \tag{2.3.38}$$

These roots have the following properties:

$$r_+(h) = [r_-(h)]^*, \quad h > 0, \tag{2.3.39}$$

$$|r_+(h)| = |r_-(h)| = 1, \quad h > 0, \tag{2.3.40}$$

$$r_+(h) = [r_-(h)]^* = e^{i\phi(h)},$$

$$\tan \phi(h) = \frac{h\sqrt{1 + \frac{h^2}{12}}}{\left(1 - \frac{h^2}{6}\right)}. \tag{2.3.41}$$

We conclude that, for $h > 0$, all solutions of Eq. (2.3.36) are oscillatory with constant amplitude.

In summary, we have seen that only the use of a discrete representation for the linear y term that is centered about the grid point t_k will give a discrete model that has oscillations with constant amplitude. Non-centered schemes allow the amplitude of the oscillations to either increase or decrease. The straightforward central difference scheme has the correct oscillatory behavior if $0 < h < 2$, while the two "symmetric" forms for y give oscillatory behavior with constant amplitude for all $h > 0$.

2.4 Logistic Differential Equation

The logistic differential equation is

$$\frac{dy}{dt} = y(1 - y). \tag{2.4.1}$$

Its exact solution can be obtained by the method of separation of variables which gives

$$y(t) = \frac{y_0}{y_0 + (1 - y_0)e^{-t}}, \tag{2.4.2}$$

where the initial condition is

$$y_0 = y(0). \tag{2.4.3}$$

Figure 2.4.1 illustrates the general nature of the various solution behaviors.

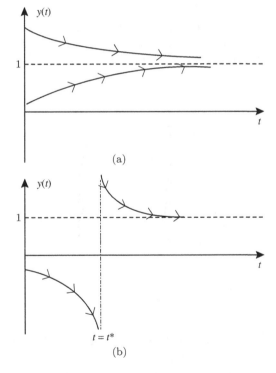

Fig. 2.4.1 Solutions of the logistic differential equation. (a) $y_0 > 0$. (b) $y_0 < 0$.

If $y_0 > 0$, then all solutions monotonically approach the stable fixed-point at $y(t) = 1$. If $y_0 < 0$, then the solution at first decreases to $-\infty$ at the singular point

$$t = t^* = \mathrm{Ln}\left[\frac{1 + |y_0|}{|y_0|}\right], \tag{2.4.4}$$

after which, for $t > t^*$, it decreases monotonically to the fixed-point at $y(t) = 1$. Note that $y(t) = 0$ is an unstable fixed-point.

Our first discrete model is constructed by using a central difference scheme for the derivative:

$$\frac{y_{k+1} - y_{k-1}}{2h} = y_k(1 - y_k). \tag{2.4.5}$$

Since Eq. (2.4.5) is a second-order difference equation, while Eq. (2.4.1) is a first-order differential equation, the value of $y_1 = y(h)$ must be determined by some procedure. We do this by use of the Euler result [7, 8, 9]

$$y_1 = y_0 + hy_0(1 - y_0). \tag{2.4.6}$$

A typical plot of the numerical solution to Eq. (2.4.5) is shown in Figure 2.4.2. This type of plot is obtained for any value of the step-size. An understanding of this result follows from a linear stability analysis of the two fixed points of Eq. (2.4.5).

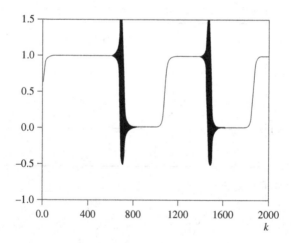

Fig. 2.4.2 Typical plot for a central difference scheme model of the logistic differential equation: $y_0 = 0.5$, $h = 0.1$. $\frac{y_{k+1} - y_{k-1}}{2h} = y_k(1 - y_k)$.

First of all, note that Eq. (2.4.5) has two constant solutions or fixed-points. They are

$$y_k = \bar{y}^{(0)} = 0, \quad y_k = \bar{y}^{(1)} = 1. \tag{2.4.7}$$

To investigate the stability of $y_k = \bar{y}^{(0)}$, we set

$$y_k = \bar{y}^{(0)} + \epsilon_k, \quad |\epsilon_k| \ll 1, \tag{2.4.8}$$

substitute this result into Eq. (2.4.5) and neglect all but the linear terms. Doing this gives

$$\frac{\epsilon_{k+1} - \epsilon_{k-1}}{2h} = \epsilon_k. \tag{2.4.9}$$

The solution to this second-order difference equation is

$$\epsilon_k = A(r_+)^k + B(r_-)^k, \tag{2.4.10}$$

where A and B are arbitrary, but, small constants; and

$$r_\pm(h) = h \pm \sqrt{1 + h^2}. \tag{2.4.11}$$

From Eq. (2.4.11), it can be concluded that the first term on the right-side of Eq. (2.4.10) is exponentially increasing, while the second term oscillates with an exponentially decreasing amplitude.

A small perturbation to the fixed-point at $\bar{y}^{(1)} = 1$ can be represented as

$$y_k = \bar{y}^{(1)} + \eta_k, \quad |\eta_k| \ll 1. \tag{2.4.12}$$

The linear perturbation equation for η_k is

$$\frac{\eta_{k+1} - \eta_{k-1}}{2h} = -\eta_k, \tag{2.4.13}$$

whose solution is

$$\eta_k = C(S_+)^k + D(S_-)^k, \tag{2.4.14}$$

where C and D are small arbitrary constants, and

$$S_\pm(h) = -h \pm \sqrt{1 + h^2}. \tag{2.4.15}$$

Thus, the first term on the right-side of Eq. (2.4.14) exponentially decreases, while the second term oscillates with an exponentially increasing amplitude.

Putting these results together, it follows that the central difference scheme has exactly the same two fixed-points as the logistic differential equation. However, while $y(t) = 0$ is (linearly) unstable and $y(t) = 1$ is (linearly) stable for the differential equation, both fixed-points are linearly unstable for the central difference scheme. The results of the linear stability analysis, as given in Eqs. (2.4.10) and (2.4.14), explain what is shown by Figure 2.4.2. For initial value y_0, such that $0 < y_0 < 1$, the values of y_k increase and exponentially approach the fixed-point $\bar{y}^{(1)} = 1$; y_k then begins to oscillate with an exponentially increasing amplitude about $\bar{y}^{(1)} = 1$ until it reaches the neighborhood of the fixed-point $\bar{y}^{(0)} = 0$. After an initial exponential decrease to $\bar{y}^{(0)} = 0$, the y_k values then begin their increase back to the fixed-point at $\bar{y}^{(1)} = 0$.

It has been shown by Yamaguti and Ushiki [10] and by Ushiki [11] that the central difference scheme allows for the existence of chaotic orbits for all positive time-steps for the logistic differential equation. Additional work on this problem has been done by other researchers including Sanz-Serna [12] and Mickens [13]. The major conclusion is that the use of the central difference scheme

$$\frac{y_{k+1} - y_{k-1}}{2h} = f(y_k), \tag{2.4.16}$$

for the scalar first-order differential equation

$$\frac{dy}{dt} = f(y) \tag{2.4.17}$$

forces all the fixed-points to become unstable [13]. Consequently, the central difference discrete derivative should never be used for this class of ordinary differential equation.

However, before leaving the use of the central difference scheme, let us consider the following discrete model for the logistic equation:

$$\frac{y_{k+1} - y_{k-1}}{2h} = y_{k-1}(1 - y_{k+1}). \tag{2.4.18}$$

Our major reason for studying this model is that an exact analytic solution exists for Eq. (2.4.18). Observe that the function

$$f(y) = y(1 - y) \tag{2.4.19}$$

is modeled locally on the lattice in Eq. (2.4.5), while it is modeled nonlocally in Eq. (2.4.18), i.e., at lattice points $k - 1$ and $k + 1$.

The substitution

$$y_k = \frac{1}{x_k}, \tag{2.4.20}$$

transforms Eq. (2.4.18) to the expression

$$x_{k+1} - \left(\frac{1}{1 + 2h}\right) x_{k-1} = \frac{2h}{1 + 2h}. \tag{2.4.21}$$

Note that Eq. (2.4.18) is a nonlinear, second-order difference equation, while Eq. (2.4.21) is a linear, inhomogeneous equation with constant coefficients. Solving Eq. (2.4.21) gives the general solution

$$x_k = 1 + [A + B(-1)^k](1 + 2h)^{k/2}, \tag{2.4.22}$$

where A and B are arbitrary constants. Therefore, y_k is

$$y_k = \frac{1}{1 + [A + B(-1)^k](1 + 2h)^{-k/2}}. \tag{2.4.23}$$

For y_0 such that $0 < y_0 < 1$, and y_1 selected such that $y_1 = y_0 + hy_0(1-y_0)$, the solutions to Eq. (2.4.23) have the structure indicated in Figure 2.4.3. Observe that the numerical solution has the general properties of the solution to the logistic differential equation, see Figure 2.4.1, except that small oscillations occur about the smooth solution.

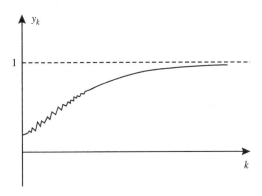

Fig. 2.4.3   A trajectory for the central difference scheme $\frac{y_{k+1} - y_{k-1}}{2h} = y_{k-1}(1 - y_{k+1})$.

The direct forward-Euler discrete model for the logistic differential equation is

$$\frac{y_{k+1} - y_k}{h} = y_k(1 - y_k). \tag{2.4.24}$$

This first-order difference equation has two constant solutions or fixed-points at $\bar{y}^{(0)} = 0$ and $\bar{y}^{(1)} = 1$. Perturbations about these fixed-points, i.e.,

$$y_k = \bar{y}^{(0)} + \epsilon_k = \epsilon_k, \quad |\epsilon_k| \ll 1, \tag{2.4.25}$$

$$y_k = \bar{y}^{(1)} + \eta_k = 1 + \eta_k, \quad |\eta_k| \ll 1, \tag{2.4.26}$$

give the following solutions for ϵ_k and η_k:

$$\epsilon_k = \epsilon_0(1 + h)^k, \tag{2.4.27}$$

$$\eta_k = \eta_0(1 - h)^k. \tag{2.4.28}$$

The expression for ϵ_k shows that $\bar{y}^{(0)}$ is unstable for all $h > 0$; thus, this discrete scheme has the same linear stability property as the differential equation for all $h > 0$. However, the linear stability properties of the fixed-point $y^{(1)}$ depend on the value of the step-size. For example:

(i) $0 < h < 1$: $\bar{y}^{(1)}$ is linearly stable; perturbations decrease exponentially.

(ii) $1 < h < 2$: $\bar{y}^{(1)}$ is linearly stable; however, the perturbations decrease exponentially with an oscillating amplitude.

(iii) $h > 2$: $\bar{y}^{(1)}$ is linearly unstable; the perturbations oscillate with an exponentially increasing amplitude.

Our conclusion is that the forward-Euler scheme gives the correct linear stability properties only if $0 < h < 1$. For this interval of step-size values, the qualitative properties of the solutions to the differential and difference equations are the same. Consequently, for $0 < h < 1$, there are no numerical instabilities.

Figure 2.4.4 presents three numerical solutions for the forward-Euler scheme given by Eq. (2.4.24). In all three cases the initial condition is $y_0 = 0.5$. The step-sizes are $h = 0.01$, 1.5 and 2.5.

Finally, it should be stated that the change of variables

$$z_k = \left(\frac{h}{1+h} \right) y_k, \quad \lambda = 1 + h, \tag{2.4.29}$$

when substituted into Eq. (2.4.24) gives the famous logistic difference equation [7, 14, 15]

$$z_{k+1} = \lambda z_k (1 - z_k). \tag{2.4.30}$$

Depending on the value of the parameter λ, this equation has a host of solutions with various periods, as well as chaotic solutions [16, 17].

Our next model of the logistic differential equation is constructed by using a forward-Euler for the first-derivative and a nonlocal expression for the function $f(y) = y(1 - y)$. This model is

$$\frac{y_{k+1} - y_k}{h} = y_k (1 - y_{k+1}). \tag{2.4.31}$$

This first-order, nonlinear difference equation can be solved exactly by using the variable change

$$y_k = \frac{1}{x_k}, \tag{2.4.32}$$

to obtain

$$x_{k+1} - \left(\frac{1}{1+h} \right) x_k = \frac{h}{1+h}, \tag{2.4.33}$$

whose general solution is

$$x_k = 1 + A(1+h)^{-k}, \tag{2.4.34}$$

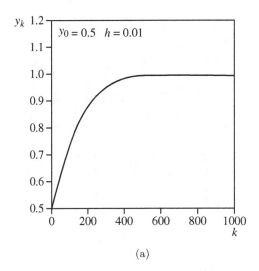

(a)

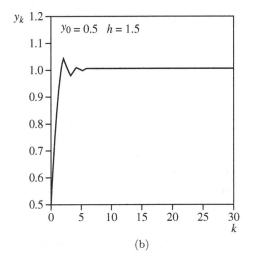

(b)

Fig. 2.4.4   The forward-Euler scheme $\frac{y_{k+1}-y_k}{h} = y_k(1-y_k)$. (a) $y_0 = 0.5$, $h = 0.01$.
(b) $y_0 = 0.5$, $h = 1.5$.

where A is an arbitrary constant. Imposing the initial condition

$$x_0 = \frac{1}{y_0}, \qquad (2.4.35)$$

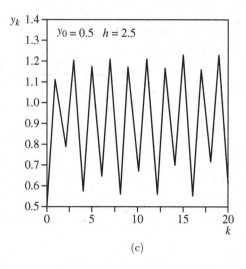

(c)

Fig. 2.4.4   The forward-Euler scheme $\frac{y_{k+1}-y_k}{h} = y_k(1-y_k)$. (c) $y_0 = 0.5$, $h = 2.4$.

gives

$$A = \frac{1 - y_0}{y_0}, \qquad (2.4.36)$$

and

$$y_k = \frac{y_0}{y_0 + (1 - y_0)(1 + h)^{-k}}. \qquad (2.4.37)$$

Examination of Eq. (2.4.37) shows that, for $h > 0$, its qualitative properties are the same as the corresponding exact solution to the logistic differential equation, namely, Eq. (2.4.2). Hence, the forward-Euler, nonlocal discrete model has no numerical instabilities for any step-size. Figure 2.4.5 gives numerical solutions using Eq. (2.4.31) for three step-sizes.

Note that Eq. (2.4.31) can be written in explicit form

$$y_{k+1} = \frac{(1 + h)y_k}{1 + hy_k}. \qquad (2.4.38)$$

Our last discrete model for the logistic differential equation is based on a second-order Runge-Kutta method [8, 9]. This technique gives for the first-order scalar equation

$$\frac{dy}{dt} = f(y), \qquad (2.4.39)$$

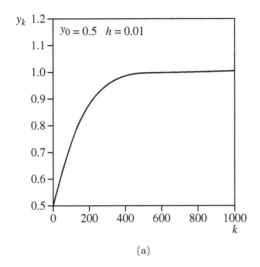

(a)

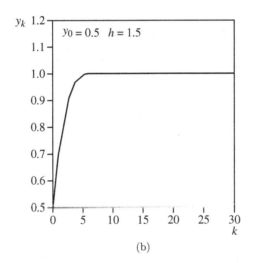

(b)

Fig. 2.4.5 Numerical solutions of $\frac{y_{k+1}-y_k}{h} = y_k(1-y_{k+1})$. (a) $y_0 = 0.5$, $h = 0.01$. (b) $y_0 = 0.5$, $h = 1.5$.

the discrete result

$$\frac{y_{k+1} - y_k}{h} = \frac{f(y_k) + f[y_k + hf(y_k)]}{2}. \qquad (2.4.40)$$

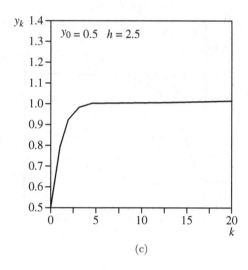

(c)

Fig. 2.4.5   Numerical solutions of $\frac{y_{k+1}-y_k}{h} = y_k(1-y_{k+1})$. (c) $y_0 = 0.5$, $h = 2.5$.

Applying this to the logistic equation, where $f(y) = y(1-y)$, gives

$$y_{k+1} = \left[1 + \frac{(2+h)h}{2}\right] y_k - \left[\frac{(2+3h+h^2)h}{2}\right] y_k^2$$

$$+ (1+h)h^2 y_k^3 - \left(\frac{h^3}{2}\right) y_k^4. \qquad (2.4.41)$$

This first-order, nonlinear difference equation has four fixed-points. They are located at

$$\bar{y}^{(0)} = 0, \quad \bar{y}^{(1)} = 1, \qquad (2.4.42)$$

$$\bar{y}^{(2,3)} = \left(\frac{1}{2h}\right)\left[(2+h) \pm \sqrt{h^2-4}\right]. \qquad (2.4.43)$$

The first two fixed-points, $\bar{y}^{(0)}$ and $\bar{y}^{(1)}$ correspond to the two fixed-points of the logistic differential equation. The other two fixed-points, $\bar{y}^{(2)}$ and $\bar{y}^{(3)}$, are spurious fixed points and are introduced by the second-order Runge-Kutta method. Note that for $h \leq 2$, the fixed-points $\bar{y}^{(2)}$ and $\bar{y}^{(3)}$ are complex conjugates of each other; while for $h \geq 2$, all fixed-points are real. Figure 2.4.6 gives a plot of all the fixed points as a function of the step-size h.

For $0 < h < 2$, there are only two real fixed-points, namely, $\bar{y}^{(0)} = 0$ and $\bar{y}^{(1)} = 1$. The first is linearly unstable and the second is linearly stable.

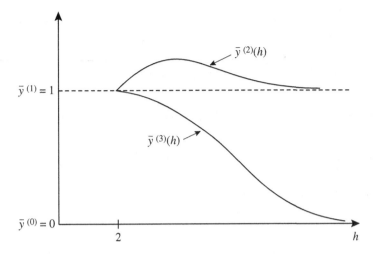

Fig. 2.4.6 Plot of the fixed-points of the 2nd-order Runge-Kutta method for the logistic differential equation. Only the spurious fixed-points depend on h.

All numerical solutions of Eq. (2.4.41), with $y_0 > 0$, thus approach $\bar{y}^{(1)}$ as $k \to \infty$. However, for $h > 2$, there exists four real fixed-points. Their order and linear stability properties are indicated below where U and S, respectively, mean linearly unstable and linearly stable:

$$\bar{y}^{(0)} < \bar{y}^{(3)}(h) < \bar{y}^{(1)} < \bar{y}^{(2)}(h)$$
$$U \qquad S \qquad U \qquad S.$$

These results and Eq. (2.4.43) predict that at a step-size of $h = 2.5$, if the initial value y_0 is selected so that $0 < y_0 < 1$, then the numerical solution of Eq. (2.4.41) will converge to the value 0.6. The validity of this prediction is shown in Figure 2.4.7(c). This figure also gives numerical solutions for several other step-sizes.

The application of the second-order Runge-Kutta method illustrates the generation of numerical instabilities that arise from the creation of additional spurious fixed-points.

Comparing the five finite difference schemes that were used to model the logistic differential equation, the nonlocal forward-Euler method clearly gave the best results. For all values of the step-size it has solutions that are in qualitative agreement with the corresponding solutions of the differential equation. The other discrete models had, for certain values of step-size, numerical instabilities.

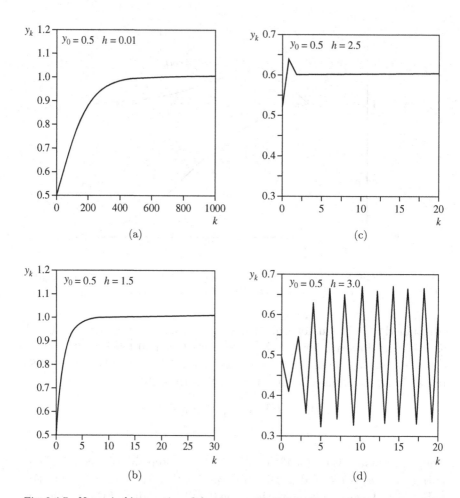

Fig. 2.4.7   Numerical integration of the logistic equation by a 2nd-order Runge-Kutta method, (a) $y_0 = 0.5$, $h = 0.01$. (b) $y_0 = 0.5$, $h = 1.5$.

Fig. 2.4.7   Numerical integration of the logistic equation by a 2nd-order Runge-Kutta method, (c) $y_0 = 0.5$, $h = 2.5$. (d) $y_0 = 0.5$, $h = 3.0$.

2.5   Unidirectional Wave Equation

The one-way or unidirectional wave equation is [10]

$$u_t + u_x = 0, \quad u(x,0) = f(x), \tag{2.5.1}$$

where the initial profile function $f(x)$ is assumed to have a first derivative. The solution to the initial value problem of Eq. (2.5.1) is

$$u(x,t) = f(x - t). \tag{2.5.2}$$

This represents a waveform moving to the right with unit velocity.

A discrete model for the unidirectional wave equation that uses forward-Euler expressions for both the time and space derivatives is

$$\frac{u_m^{k+1} - u_m^k}{\Delta t} + \frac{u_{m+1}^k - u_m^k}{\Delta x} = 0, \tag{2.5.3}$$

where

$$t_k = (\Delta t)k, \quad x_m = (\Delta x)m, \tag{2.5.4}$$

are the discrete time and space variables. Define β to be

$$\beta = \frac{\Delta t}{\Delta x}. \tag{2.5.5}$$

Using this definition, Eq. (2.5.3) takes the form

$$u_m^{k+1} + \beta u_{m+1}^k - (1 + \beta)u_m^k = 0. \tag{2.5.6}$$

The method of separation of variables can be used to obtain particular solutions to Eq. (2.5.6). Assume that u_m^k can be written as [7]

$$u_m^k = C^k D_m. \tag{2.5.7}$$

(Note that C^k is a function of the discrete variable k and does not mean "C" raised to the k-th power. The same statement applies to D_m.) Substitution of this result into Eq. (2.5.6) gives

$$C^{k+1} D_m - C^k [\beta D_{m+1} - (1 + \beta D_m] = 0, \tag{2.5.8}$$

and

$$\frac{C^{k+1}}{C^k} + \frac{\beta D_{m+1} - (1 + \beta)D_m}{D_m} = 0. \tag{2.5.9}$$

Since the first term depends only on k, while the second term depends only on m, each term must be constant. Denoting the "separation constant" by α, we obtain

$$C^{k+1} = \alpha C^k, \tag{2.5.10}$$

$$\beta D_{m+1} - (1 + \beta)D_m = -\alpha D_m. \tag{2.5.11}$$

These equations have the respective solutions

$$C^k = A(\alpha)\alpha^k, \tag{2.5.12}$$

$$D_m = B(\alpha) \left(\frac{1 + \alpha + \beta}{\beta} \right)^m, \tag{2.5.13}$$

where $A(\alpha)$ and $B(\alpha)$ are "constants." Therefore, a particular solution to Eq. (2.5.6) is

$$u_m^k(\alpha, \beta) = E(\alpha)\alpha^k \left(\frac{1 + \alpha + \beta}{\beta}\right)^m, \qquad (2.5.14)$$

$$E(\alpha) = A(\alpha)B(\alpha). \qquad (2.5.15)$$

The Eq. (2.5.14) can also be written in the form

$$u_m^k(\alpha, \beta) = E(\alpha)\alpha^{(k-m)} \left[\frac{(1 + \alpha + \beta)\alpha}{\beta}\right]^m. \qquad (2.5.16)$$

If β is chosen to be

$$\beta = 1. \qquad (2.5.17)$$

and if a sum/integral is done over α, then the following general solution is obtained [7]

$$
\begin{aligned}
u_m^k &= \sum_\alpha u_m^k(\alpha, 1) \\
&= \sum_\alpha \bar{\alpha}^{(t_k - x_m)} (2 + \alpha)^m, \quad \bar{\alpha} = \alpha^{1/\Delta x}, \\
&\neq g(x_m - t_k),
\end{aligned}
\qquad (2.5.18)
$$

where $g(z)$ is an arbitrary function of z. Our general conclusion is that the finite difference scheme, of Eq. (2.5.3), does not have solutions that correspond exactly to those of the unidirectional wave equation.

For a second model, let us replace the time and space derivatives by, respectively, forward-Euler and central difference expressions. Doing this gives

$$\frac{u_m^{k+1} - u_m^k}{\Delta t} + \frac{u_{m+1}^k - u_{m-1}^k}{\Delta x} = 0, \qquad (2.5.19)$$

and, upon rearranging, the equation

$$u_m^{k+1} + \beta u_{m+1}^k - u_m^k - \beta u_{m-1}^k = 0. \qquad (2.5.20)$$

Assuming a particular solution, $u_m^k = C^k D_m$, we obtain

$$C^{k+1} D_m - C^k [D_{m+1} - D_m - \beta D_{m-1}] = 0. \qquad (2.5.21)$$

Let ζ be the separation constant. The equations for C^k and D_m are

$$C^{k+1} = \zeta C^k, \qquad (2.5.22)$$

$$\beta D_{m+1} + (\zeta - 1)D_m - \beta D_{m-1} = 0. \qquad (2.5.23)$$

These difference equations have the solutions

$$C^k = A(\zeta)\zeta^k \tag{2.5.24}$$

$$D_m = B_1(\zeta)[r_+(\beta,\zeta)]^k + B_2(\zeta)[r_-(\beta,\zeta)]^k, \tag{2.5.25}$$

where $r_+(\beta,\zeta)$ and $r_-(\beta,\zeta)$ are roots to the characteristic equation [7]

$$\beta r^2 + (\zeta - 1)r - \beta = 0. \tag{2.5.26}$$

Therefore,

$$u_m^k(\beta,\zeta) = \{H_1(\zeta)[r_+(\beta,\zeta)]^m + H_2(\zeta)[r_-(\beta,\zeta)]^m\zeta_k, \tag{2.5.27}$$

and a general solution is

$$u_m^k(\beta) = \sum_\zeta u_m^k(\beta,\zeta) \neq g(x_m - t_k). \tag{2.5.28}$$

Again, we conclude that Eq. (2.5.19) does not provide a good discrete model for the unidirectional wave equation.

Finally, consider a discrete model for which the time and space derivatives are given, respectively, by forward- and backward-Euler expressions. For this case, we have

$$\frac{u_m^{k+1} - u_m^k}{\Delta t} + \frac{u_m^k - u_{m-1}^k}{\Delta} = 0 \tag{2.5.29}$$

and

$$u_m^{k+1} + (\beta - 1)u_m^k - \beta u_{m-1}^k = 0. \tag{2.5.30}$$

The separation of variables equations are

$$C^{k+1} = \gamma C^k, \tag{2.5.31}$$

$$(\beta + \gamma - 1)D_m - \beta D_{m-1} = 0, \tag{2.5.32}$$

where γ is the separation constant. The solutions to these equations are

$$C^k = A(\gamma)\gamma^k, \tag{2.5.33}$$

$$D_m = B(\gamma)\left(\frac{1}{\beta + \gamma - 1}\right)^m. \tag{2.5.34}$$

With $G(\gamma) = A(\gamma)B(\gamma)$, the general solution is

$$u_m^k(\beta) = \sum_\gamma G(\gamma)\gamma^k(\beta + \gamma - 1)^{-m}. \tag{2.5.35}$$

Note that for general β, we have

$$u_m^k(\beta) \neq g(x_m - t_k). \tag{2.5.36}$$

However, for $\beta = 1$ or $\Delta t = \Delta x$, Eq. (2.5.35) becomes

$$u_m^k(1) = \sum_\gamma G(\gamma)\gamma^{(k-m)} = g(x_m - t_k). \tag{2.5.37}$$

Consequently, if $\beta = 1$, the discrete model of Eq. (2.5.29) has solutions that are exactly equal to the solution of the unidirectional wave equation on the computational lattice. Under this condition, Eq. (2.5.29) reduces to the simpler form

$$u_m^{k+1} = u_{m-1}^k, \quad \Delta t = \Delta x. \tag{2.5.38}$$

2.6   Burgers' Equation

The full Burgers' equation is [18, 19]

$$u_t + u u_x = e u_{xx}, \tag{2.6.1}$$

where e is related to the reciprocal of the Reynolds number [18]. For the present study, we consider the case for which $e = 0$, i.e.,

$$u_t + u u_x = 0. \tag{2.6.2}$$

This first-order, nonlinear partial differential equation has no exact explicit solution for the initial value problem $u(x, 0) = f(x)$ where $f(z)$ has a first derivative [18]. However, it does have a particular solution that can be obtained by the method of separation of variables. Assume

$$u(x, t) = C(t)D(x), \tag{2.6.3}$$

and substitute this into Eq. (2.6.2) to obtain

$$\frac{dC}{dt}D + CDC\frac{dD}{dx} = 0, \tag{2.6.4}$$

or

$$\frac{1}{C^2}\frac{dC}{dt} + \frac{dD}{dx} = 0. \tag{2.6.5}$$

Denoting the separation constant by a, these ordinary differential equations have the solutions

$$C(t) = \frac{1}{at + d}, \tag{2.6.6}$$

$$D(x) = ax + b, \tag{2.6.7}$$

where b and d are arbitrary integration constants. Hence, a particular, rational solution of the Burgers' equation is

$$u(x, t) = \frac{ax + b}{at + d}. \tag{2.6.8}$$

Now consider a discrete model of the Burgers' equation that uses forward-Eulers for both the time and space derivatives:

$$\frac{u_m^{k+1} - u_m^k}{\Delta t} + u_m^k \left(\frac{u_{m+1}^k - u_m^k}{\Delta x} \right) = 0. \tag{2.6.9}$$

For $\beta = \Delta t / \Delta x$, this equation takes the form

$$u_m^{k+1} - u_m^k + \beta u_m^k u_{m+1}^k - \beta (u_m^k)^2 = 0. \tag{2.6.10}$$

Let $u_m^k = C^k D_m$, then the equations satisfied by C^k and D_m are

$$C^{k+1} = C^k - \alpha (C^k)^2, \tag{2.6.11}$$

$$D_{m+1} - D_m = \frac{\alpha}{\beta}, \tag{2.6.12}$$

where α is the separation constant. The general solution to Eq. (2.6.12) is

$$D_m = \left(\frac{\alpha}{\beta} \right) m + b_1, \tag{2.6.13}$$

where β_1 is an arbitrary constant. If we now make the identifications

$$\frac{\alpha}{\beta} = a(\Delta x), \quad b_1 = b, \tag{2.6.14}$$

then Eq. (2.6.13) becomes

$$D_m = a x_m + b, \tag{2.6.15}$$

which is the discrete version of Eq. (2.6.7). However, Eq. (2.6.11) is the logistic difference equation for which no general solution exists in terms of a finite sum of elementary functions [1, 7]. Therefore, the discrete version of Eq. (2.6.6) is not a solution of Eq. (2.6.11). Our conclusion is that Eq. (2.6.9) will have solutions that do not correspond to any solution of the Burgers' partial differential equation; consequently, this scheme has numerical instabilities.

Similar results are obtained for the discrete model

$$\frac{u_m^{k+1} - u_m^k}{\Delta t} + u_m^k \left(\frac{u_m^k - u_{m-1}^k}{\Delta x} \right) = 0 \tag{2.6.16}$$

for which the separation of variables equations are

$$C^{k+1} = C^k - \alpha \beta (C^k)^2, \tag{2.6.17}$$

$$D_{m+1} - D_m = \frac{\alpha}{\beta}. \tag{2.6.18}$$

2.7 Summary

What have we learned from the study of various discrete models of several linear and nonlinear differential equations? The results stated below are based not only on those equations investigated in this chapter, but also on other differential equations and their associated finite difference models [20–22].

First, if the order of the finite difference scheme is greater than the order of the differential equation, then numerical instabilities will certainly occur for all step-sizes. This type of behavior is illustrated by the use of the central difference scheme for the first derivative in both the decay and logistic equations. Mathematically, this type of instability occurs because the higher-order difference equation has a larger set of general solutions than the corresponding differential equation. For example, the linear decay equation has but one solution. However, a discrete model that uses the central difference scheme has two linearly independent solutions since it is of second-order.

Second, most discrete models require restrictions on the step-size to ensure that numerical instabilities do not occur. All forward-Euler type schemes and their generalizations, such as Runge-Kutta methods, have this property.

Third, for many ordinary differential equations, a linear stability analysis of the fixed-points allows a determination of when numerical instabilities occur.

Fourth, the use of nonlocal representations of non-derivative terms can often eliminate numerical instabilities, as was the case for the logistic differential equation with a forward-Euler discrete derivative. In some instances, for example, in the application of the central difference scheme to the logistic equation, a nonlocal model gave solutions that followed rather closely the trajectories of the solution to the differential equation except for small oscillations.

Fifth, for discrete models of partial differential equations, the use of forward- or backward-Euler schemes for the first-derivatives can have a significant impact on the solution behaviors of the equations.

We now demonstrate that, in general, numerical instabilities always occur in the discrete modeling of ordinary differential equations if one uses either the central difference or the forward-Euler schemes provided the non-derivative terms are modeled locally on the computational grid. For our purposes, it is sufficient to prove this for the scalar equation

$$\frac{dy}{dt} = f(y), \tag{2.7.1}$$

where

$$f(y) = 0, \tag{2.7.2}$$

is assumed to have only simple zeros. For this autonomous, first-order differential equation, numerical instabilities will occur whenever the linear stability properties of any of the fixed-points for the discrete model differs from those of the differential equation [13, 23, 24, 25].

The fixed-points or constant solutions of Eq. (2.7.1) are solutions to the equation

$$f(\bar{y}) = 0. \tag{2.7.3}$$

Denote these zeros by $\{\bar{y}^{(i)}\}$, where $i = 1, 2, \dots, I$. Note that I may be unbounded. Now, define R_i as follows

$$R_i \equiv \frac{df[\bar{y}^{(i)}]}{dy}. \tag{2.7.4}$$

The application of linear stability analysis to the i-th fixed-point gives the following result [26]:

(i) If $R_i > 0$, the fixed-point $y(t) = \bar{y}^{(i)}$ is linearly unstable.
(ii) If $R_i < 0$, the fixed-point $y(t) = \bar{y}^{(i)}$ is linearly stable.

Now construct a central difference discrete model for Eq. (2.7.1), i.e.,

$$\frac{y_{k+1} - y_{k-1}}{2h} = f(y_k). \tag{2.7.5}$$

For small perturbations, ϵ_k, about the fixed-point $\bar{y}^{(i)}$, we have

$$y_k = \bar{y}^{(i)} + \epsilon_k. \tag{2.7.6}$$

If Eq. (2.7.6) is substituted into Eq. (2.7.5) and only linear terms are kept, then we obtain

$$\frac{\epsilon_{k+1} - \epsilon_{k-1}}{2h} = R_i \epsilon_k. \tag{2.7.7}$$

An examination of the characteristic equation for Eq. (2.7.7)

$$r^2 - (2hR_i)r - 1 = 0 \tag{2.7.8}$$

shows that one root is always larger than one in magnitude. In fact,

$$r_\pm = hR_i \pm \sqrt{1 + h^2 R_i^2}. \tag{2.7.9}$$

Since,

$$\epsilon_k = A(r_+)^k + B(r_-)^k, \qquad (2.7.10)$$

where A and B are arbitrary, but small constants, we must conclude that the fixed-point at $y_k = \bar{y}^{(i)}$ is linearly unstable. However, if $R_i < 0$, then the corresponding fixed-point of the differential equation is stable. Therefore, the use of the central difference scheme of Eq. (2.7.5) leads to a discrete model of Eq. (2.7.1) for which all the fixed-points are linearly unstable. This means that the central scheme has numerical instabilities for all $h > 0$. As stated previously, the main reason for the occurrence of numerical instabilities in this case is that the order of the finite difference equation is larger than the order of the corresponding differential equation.

Let us now investigate the linear stability properties of the fixed-points for the forward-Euler scheme for Eq. (2.7.1). It is given by the following expression

$$\frac{y_{k+1} - y_k}{h} = f(y_k). \qquad (2.7.11)$$

A perturbation of the i-th fixed-point, as given by Eq. (2.7.6), leads to the perturbation equation

$$\frac{\epsilon_{k+1} - \epsilon_k}{h} = R_i \epsilon_k, \qquad (2.7.12)$$

or

$$\epsilon_{k+1} = (1 + hR_i)\epsilon_k, \qquad (2.7.13)$$

which has the solution

$$\epsilon_k = \epsilon_0 (1 + hR_i)^k. \qquad (2.7.14)$$

Detailed study of Eq. (2.7.14) gives the following results:

(i) For $R_i > 0$, the fixed-point $\bar{y}^{(i)}$ is linearly unstable for both the differential Eq. (2.7.1) and the difference Eq. (2.7.11) for $h > 0$.

(ii) For $R_i < 0$, which corresponds to a linearly stable fixed-point for the differential Eq. (2.7.1), the fixed-point of the discrete model, namely Eq. (2.7.11), has the properties:

$$0 < h < \frac{2}{|R_i|}, \quad y_k = \bar{y}^{(i)} \text{ is linearly stable};$$

$$h \geq \frac{2}{|R_i|}, \quad y_k = \bar{y}^{(i)} \text{ is linearly unstable}.$$

Consequently, we conclude that the forward-Euler scheme and the differential equation will have corresponding fixed-points with the same linear stability properties only if there is a limitation on the step-size h, i.e.,

$$0 < h < h^* = \frac{2}{R^*}, \tag{2.7.15}$$

where

$$R^* = \text{Max}\{|R_i|; i = 1, 2, \ldots, I\}. \tag{2.7.16}$$

Numerical instabilities will occur whenever $h > h^*$. This type of numerical instability will be called a threshold instability.

Note that for the central difference scheme $h^* = 0$, i.e., numerical instabilities occur for all $h > 0$.

The previous two finite difference methods were explicit schemes. We now investigate the properties of an implicit discrete model for Eq. (2.7.1), the backward-Euler scheme. It is given by the expression

$$\frac{y_{k+1} - y_k}{h} = f(y_{k+1}). \tag{2.7.17}$$

For small perturbations about the fixed-point at $y_k = \bar{y}^{(i)}$ the equation for ϵ_k is

$$\frac{\epsilon_{k+1} - \epsilon_k}{h} = R_i \epsilon_{k+1}, \tag{2.7.18}$$

or

$$\epsilon_{k+1} = \left(\frac{1}{1 - hR_i}\right)\epsilon_k, \tag{2.7.19}$$

which has the solution

$$\epsilon_k = \epsilon_0 \left(\frac{1}{1 - hR_i}\right)^k. \tag{2.7.20}$$

Inspection of Eq. (2.7.20) leads to the following conclusions:

(i) For $R_i < 0$, the fixed-point of Eq. (2.7.17) is linearly stable for all $h > 0$. Thus, the stability properties of the finite difference scheme and the differential equation are the same.

(ii) For $R_i > 0$, the finite difference scheme is linearly unstable for

$$0 < h \leq \frac{2}{R_i}, \tag{2.7.21}$$

but, is linearly stable for

$$h > \frac{2}{R_i}. \tag{2.7.22}$$

Note that for
$$h > \frac{2}{R}, \quad \bar{R} = \text{Min}\{|R_i|; \ i = 1, 2, \ldots, I\}, \qquad (2.7.23)$$
all the fixed-points of this implicit scheme are linearly stable. This phenomena is called super-stability by Dahlquist et al. [27] and has been investigated by Lorenz [28], Dieci and Estep [29], and Corless et al. [24]. This phenomena is of great interest since, for systems of ordinary differential equations, there exist discrete models that produce solutions that are not chaotic even though the differential equations themselves are known to have chaotic behavior. This result is the "natural complement of computational chaos" (Corless et al. [24]) or numerical instabilities that can arise when certain finite difference schemes are used to construct discrete models of ordinary differential equations. Above, we have shown that super-stability can also occur in the backward-Euler scheme for a single scalar equation.

The next chapter will be devoted to the study of nonstandard finite difference schemes and how they can be used to eliminate the elementary forms of numerical instabilities as shown to exist in the present chapter.

References

1. G. Iooss, *Bifurcation of Maps and Applications* (North-Holland, Amsterdam, 1979).
2. V. Arnold, *Geometrical Methods in the Theory of Ordinary Differential Equations* (Springer-Verlag, New York, 1983).
3. G. Iooss and M. Adelmeyer, *Topics in Bifurcation Theory and Applications* (World Scientific, Singapore, 1992).
4. F. B. Hilderbrand, *Finite-Difference Equations and Simulations* (Prentice-Hall; Englewood Cliffs, NJ; 1968).
5. V. D. Barger and M. G. Olsson, *Classical Mechanics: A Modern Perspetive* (McGraw-Hill, New York, 1973).
6. R. E. Mickens, *Nonlinear Oscillations* (Cambridge University Press, New York, 1981).
7. R. E. Mickens, *Difference Equations: Theory and Applications* (Van Nostrand Reinhold, New York, 1990).
8. M. K. Jain, *Numerical Solution of Differential Equations* (Wiley, New York, 2nd edition, 1984).
9. J. M. Ortega and W. G. Poole, Jr., *Numerical Methods for Differential Equations* (Pitman; Mashfield, MA; 1981).
10. M. Yamaguti and S. Ushiki, *Physica* 3D, 618–626 (1981). Chaos in numerical analysis of ordinary differential equations.

11. S. Ushiki, *Physica* 4D, 407–424 (1982). Central difference scheme and chaos.

12. J. M. Sanz-Serna, *SIAM Journal of Scientific and Statistical Computing* 6, 923–938 (1985). Studies in numerical nonlinear instability I. Why do leapfrog schemes go unstable?

13. R. E. Mickens, *Dynamic Systems and Applications* 1, 329–340 (1992). Finite difference schemes having the correct linear stability properties for all finite step-sizes II.

14. R. M. May, *Nature* **261**, 459–467 (1976). Simple mathematical models with very complicated dynamics.

15. P. Collet and J.-P. Eckmann, *Iterated Maps of the Interval as Dynamical Systems* (Birkhauser, Boston, 1980).

16. T. L. and J. Yorke, *American Mathematical Monthly* **82**, 985–992 (1975). Period-3 implies chaos.

17. R. L. Devaney, *An Introduction to Chaotic Dynamical Systems* (Benjamin/Cummings; Menlo Park, CA; 1986).

18. G. B. Whitham, *Linear and Nonlinear Waves* (Wiley-Interscience, New York, 1974).

19. J. M. Burgers, *Advanced in Applied Mechanics* 1, 171–199 (1948). A mathematical model illustrating the theory of turbulence.

20. R. E. Mickens, Difference equation models of differential equations having zero local truncation errors, in *Differential Equations*, I. W. Knowles and R. T. Lewis, editors (North-Holland, Amsterdam, 1984), pp. 445–449.

21. R. E. Mickens, Mathematical modeling of differential equations by difference equations, in *Computational Acoustics: Wave Propagation*, D. Lee et al., editors (Elsevier Science Publications B. V., Amsterdam, 1988), pp. 387–393.

22. R. E. Mickens, *Numerical Methods for Partial Differential Equations* 5, 313–325 (1989). Exact solutions to a finite-difference model for a nonlinear reaction-advection equation: Implications for numerical analysis.

23. R. E. Mickens, Runge-Kutta schemes and numerical instabilities: The logistic equation, in *Differential Equations and Mathematical Physics*, I. Knowles and Y. Saito, editors (Springer-Verlag, Berlin, 1987), pp. 337–341.

24. R. M. Corless, C. Essex and M. A. H. Nerenberg, *Physics Letters* **A157**, 27–36 (1991). Numerical methods can suppress chaos.

25. A. Iserles, A. T. Peplow and A. M. Stuart, *SIAM Journal of Numerical Analysis* **28**, 1723–1751 (1991). A unified approach to spurious solutions introduced by time discretization. Part I: Basic theory.

26. M. Sever, *Ordinary Differential Equations* (Boole Press, Dublin, 1987), pp. 101–103.

27. G. Dahlquist, L. Edsberg, G. Skollermo, and G. Soderlind, Are the numerical methods and software satisfactory for chemical kinetics, in *Numerical Integration of Differential Equations and Large Linear Systems*, J. Hinze, editor (Springer-Verlag, Berlin, 1982), pp. 149–164.

28. E. N. Lorentz, *Physica* **D35**, 299–317 (1989). Computational chaos – A preclude to computational instability.

29. L. Dieci and D. Estep, Georgia Institute of Technology, Tech. Rep. Math. 050290-039 (1990). Some stability aspects of schemes for the adaptive integration of stiff initial value problems.

Chapter 3

Nonstandard Finite Difference Schemes

3.1 Introduction

This chapter provides background information to understand the general rules of Mickens [1] for the construction of nonstandard finite difference schemes for differential equations. First, the concept of an exact difference scheme is introduced and defined. Second, a theorem is stated and proved that all ordinary differential equations have a unique exact difference scheme. The major consequence of this result is that such schemes do not allow numerical instabilities to occur. Third, using this theorem, exact difference schemes are constructed for a variety of both ordinary and partial differential equations. From these results are formulated a set of modeling rules for the construction of nonstandard finite difference schemes. Fourth, the notion of best difference schemes is defined and its use in the actual construction of finite difference schemes is illustrated by several examples.

Before proceeding, we would like to make several comments related to the discrete modeling of the scalar ordinary differential equation

$$\frac{dy}{dt} = f(y, \lambda), \tag{3.1.1}$$

where λ is an n-parameter vector. The most general finite difference model for Eq. (3.1.1) that is of first-order in the discrete derivative takes the following form

$$\frac{y_{k+1} - y_k}{\phi(h, \lambda)} = F(y_k, y_{k+1}, \lambda, h). \tag{3.1.2}$$

The discrete derivative, on the left-side, is a generalization [2] of that which is normally used, namely [3],

$$\frac{dy}{dt} \rightarrow \frac{y_{k+1} - y_k}{h}. \tag{3.1.3}$$

From Eq. (3.1.2), we have

$$\frac{dy}{dt} \to \frac{y_{k+1} - y_k}{\phi(h, \lambda)}, \tag{3.1.4}$$

where the *denominator function* $\phi(h, \lambda)$ has the property [2]

$$\phi(h, \lambda) = h + O(h^2),$$
$$\lambda = \text{fixed}, \quad h \to 0. \tag{3.1.5}$$

This form for the discrete derivative is based on the traditional definition of the derivative which can be generalized as follows:

$$\frac{dy}{dt} = \lim_{h \to 0} \frac{y[t + \psi_1(h)] - y_k}{\psi_2(h)}, \tag{3.1.6}$$

where

$$\psi_i(h) = h + O(h^2), \quad h \to 0; \quad i = 1, 2. \tag{3.1.7}$$

Examples of functions $\psi(h)$ that satisfy this condition are

$$\psi = h \begin{cases} h_i \\ \sin(h), \\ e^h - 1 \\ 1 - e^{-h} \\ \frac{1 - e^{-\lambda h}}{\lambda}, \\ \text{etc.} \end{cases}$$

Note that in taking the $\text{Lim}\, h \to 0$ to obtain the derivative, the use of any of these $\psi(h)$ will lead to the usual result for the first derivative

$$\frac{dy}{dt} = \lim_{h \to 0} \frac{[t + \psi_1(h)] - y(t)}{\psi_2(h)} = \lim_{h \to 0} \frac{y(t + h) - y(t)}{h}. \tag{3.1.8}$$

However, for h finite, these discrete derivatives will differ greatly from those conventionally given in the literature, such as Eq. (3.1.3). This fact not only allows for the construction of a larger class of finite difference models, but also provides for more ambiguity in the modeling process.

3.2   Exact Finite Difference Schemes

We consider only first-order, scalar ordinary differential equations in this section. However, the results can be easily generalized to coupled systems of first-order ordinary differential equations.

It should be acknowledged that the early work of Potts [4] played a fundamental role in interesting the author in the concept of exact finite difference schemes.

Consider the general first-order differential equation

$$\frac{dy}{dt} = f(y, t, \lambda), \quad y(t_0) = y_0, \tag{3.2.1}$$

where $f(y, t, \lambda)$ is such that Eq. (3.2.1) has a unique solution over the interval, $0 \le t < T$ [5, 6] and for λ in the interval $\lambda_1 \le \lambda \le \lambda_2$. (For dynamical systems of interest, in general, $T = \infty$, i.e., the solution exists for all time.) This solution can be written as

$$y(t) = \phi(\lambda, y_0, t_0, t), \tag{3.2.2}$$

with

$$\phi(\lambda, y_0, t_0, t_0) = y_0. \tag{3.2.3}$$

Now consider a discrete model of Eq. (3.2.1)

$$y_{k+1} = g(\lambda, h, y_k, t_k), \quad t_k = hk. \tag{3.2.4}$$

Its solution can be expressed in the form

$$y_k = \psi(\lambda, h, y_0, t_0, t_k), \tag{3.2.5}$$

with

$$\psi(\lambda, h, y_0, t_0, t_0) = y_0. \tag{3.2.6}$$

Definition 1. Equations (3.2.1) and (3.2.4) are said to have the *same general solution* if and only if

$$y_k = y(t_k) \tag{3.2.7}$$

for arbitrary values of h.

Definition 2. An *exact difference scheme* is one for which the solution to the difference equation has the same general solution as the associated differential equation.

These definitions lead to the following result:

Theorem. *The differential equation*

$$\frac{dy}{dt} = f(y, t, \lambda), \quad y(t_0) = y_0, \tag{3.2.8}$$

has an exact finite difference scheme given by the expression

$$y_{k+1} = \phi[\lambda, y_k, t_k, t_{k+1}], \qquad (3.2.9)$$

where ϕ is that of Eq. (3.2.2).

Proof [7]. The group property [5, 6] of the solutions to Eq. (3.2.8) gives

$$y(t + h) = \phi[\lambda, y(t), t, t + h]. \qquad (3.2.10)$$

If we now make the identifications

$$t \to t_k, \quad y(t) \to y_k, \qquad (3.2.11)$$

then Eq. (3.2.10) becomes

$$y_{k+1} = \phi(\lambda, y_k, t_k, t_{k+1}). \qquad (3.2.12)$$

This is the required ordinary difference equation that has the same general solution as Eq. (3.2.8). □

Comments. (i) If all solutions of Eq. (3.2.8) exist for all time, i.e., $T = \infty$, then Eq. (3.2.10) holds for all t and h. Otherwise, the relation is assumed to hold whenever the right-side of Eq. (3.2.10) is well defined.

(ii) The theorem is only an existence theorem. It basically says that if an ordinary differential equation has a solution, then an exact finite difference scheme exists. In general, no guidance is given as to how to actually construct such a scheme.

(iii) A major implication of the theorem is that the solution of the difference equation is exactly equal to the solution of the ordinary differential equation on the computational grid for fixed, but, arbitrary step-size h.

(iv) The theorem can be easily generalized to systems of coupled, first-order ordinary differential equations.

The question now arises as to whether exact difference schemes exist for partial differential equations. For an arbitrary partial differential equation the answer is (probably) no. This negative result is a consequence of the fact that given an arbitrary partial differential equation there exists no clear, unambiguous accepted definition of a general solution to the equation [8, 9]. However, we should expect that certain classes of partial differential equations will have exact difference models. Note that in this case some type of functional relation should exist between the various (space and time) step-sizes.

The discovery of exact discrete models for particular ordinary and partial differential equations is of great importance, primarily because it allows us to gain insights into the better construction of finite difference schemes. They also provide the computational investigator with useful benchmarks for comparison with the standard procedures.

3.3 Examples of Exact Schemes

In this section, we will use the theorem of the last section "in reverse" to construct exact difference schemes for several ordinary and partial differential equations for which exact general solutions are explicitly known. These schemes have the property that their solutions do not have numerical instabilities.

However, before proceeding, it should be indicated that given a set of linearly independent functions

$$\{y^{(i)}(t)\}; \quad i = 1, 2, \ldots, N, \tag{3.3.1}$$

it is always possible to construct an N-th order linear difference equation that has the corresponding discrete functions as solutions [10]. For let

$$y_k^{(i)} = y^{(i)}(t_k), \quad t_k = (\Delta t)k = hk; \tag{3.3.2}$$

then the following determinant gives the required difference equation

$$\begin{vmatrix} y_k & y_k^{(1)} & y_k^{(2)} & \cdots & y_k^{(n)} \\ y_{k+1} & y_{k+1}^{(1)} & y_{k+1}^{(2)} & \cdots & y_{k+1}^{(n)} \\ \vdots & \vdots & \vdots & \vdots \\ y_{k+n} & y_{k+n}^{(1)} & y_{k+n}^{(2)} & \cdots & y_{k+n}^{(n)} \end{vmatrix} = 0. \tag{3.3.3}$$

As a first example to illustrate this procedure, consider the single function

$$y^{(1)}(t) = e^{-\lambda t}. \tag{3.3.4}$$

This is (with an arbitrary multiplicative constant) the general solution to the first-order differential equation

$$\frac{dy}{dt} = -\lambda y. \tag{3.3.5}$$

The corresponding difference equation is

$$\begin{vmatrix} y_k & y_k^{(1)} \\ y_{k+1} & y_{k+1}^{(1)} \end{vmatrix} = \begin{vmatrix} y_k & e^{-\lambda hk} \\ y_{k+1} & e^{-\lambda h(k+1)} \end{vmatrix} = e^{-\lambda hk} \begin{vmatrix} y_k & 1 \\ y_{k+1} & e^{-\lambda h} \end{vmatrix}$$

$$= e^{-\lambda hk} \left[e^{-\lambda h} y_k - y_{k+1} \right] = 0, \tag{3.3.6}$$

or

$$y_{k+1} = e^{-\lambda h} y_k. \tag{3.3.7}$$

This is the exact difference equation corresponding to Eq. (3.3.5). However, a more instructive form can be obtained by carrying out the following manipulations:

$$y_{k+1} - y_k = (e^{-\lambda h} - 1)y_k = -\lambda \left(\frac{1 - e^{-\lambda h}}{\lambda} \right) y_k, \qquad (3.3.8)$$

and finally,

$$\frac{y_{k+1} - y_k}{\left(\frac{1-e^{-\lambda h}}{\lambda} \right)} = -\lambda y_k. \qquad (3.3.9)$$

Note that the standard forward-Euler scheme for this differential equation is

$$\frac{y_{k+1} - y_k}{h} = -\lambda y_k. \qquad (3.3.10)$$

For a second example, consider the harmonic oscillator differential equation

$$\frac{d^2y}{dt^2} + \omega^2 y = 0, \qquad (3.3.11)$$

where ω is a real constant. The two linearly independent solutions are

$$y^{(1)}(t) = \cos(\omega t), \quad y^{(2)}(t) = \sin(\omega t), \qquad (3.3.12)$$

or in complex form

$$y^{(1)}(t) = e^{i\omega t}, \quad y^{(2)}(t) = e^{-i\omega t}. \qquad (3.3.13)$$

Therefore,

$$\begin{vmatrix} y_k & e^{i\omega hk} & e^{-i\omega hk} \\ y_{k+1} & e^{i\omega h(k+1)} & e^{-i\omega h(k+1)} \\ y_{k+2} & e^{i\omega h(k+2)} & e^{-i\omega h(k+2)} \end{vmatrix} = 0 \qquad (3.3.14)$$

and

$$y_{k+2} - [2\cos(\omega h)]y_{k+1} + y_k = 0. \qquad (3.3.15)$$

Shifting downward the index k by one unit and using the identity

$$2\cos(\omega h) = 2 - 4\sin^2\left(\frac{\omega h}{2} \right). \qquad (3.3.16)$$

Eq. (3.3.15) can be put in the form

$$\frac{y_{k+1} - 2y_k + y_{k-1}}{\left(\frac{4}{\omega^2} \right) \sin^2 \left(\frac{h\omega}{2} \right)} + \omega^2 y_k = 0. \qquad (3.3.17)$$

This is the exact finite difference scheme for Eq. (3.3.11) and should be compared to the standard central difference model of the harmonic oscillator differential equation

$$\frac{y_{k+1} - 2y_k + y_{k-1}}{h^2} + \omega^2 y_k = 0. \tag{3.3.18}$$

For nonlinear differential equations, the above procedure cannot be used to construct exact finite difference schemes. A procedure based on the theorem of the previous section must be used. The following outlines the steps to be applied:

(i) Consider a system of N coupled, first-order, ordinary differential equations

$$\frac{dY}{dt} = F(Y, t, \lambda), \quad Y(t_0) = Y_0, \tag{3.3.19}$$

where Y, F are N-dimensional column vectors whose i-th components are

$$(Y)_i = y^{(i)}(t), \tag{3.3.20}$$

$$(F)_i = f^{(i)}[y^{(1)}, y^{(2)}, \ldots, y^{(N)}; t, \lambda]. \tag{3.3.21}$$

(ii) Denote the general solution to Eq. (3.3.19) by

$$Y(t) = \Phi(\lambda, y_0, t_0, t) \tag{3.3.22}$$

where

$$y^{(i)}(t) = \phi^{(i)}[\lambda, y_0^{(1)}, y_0^{(2)}, \ldots, y_0^{(N)}, t_0, t]. \tag{3.3.23}$$

(iii) The exact difference equation corresponding to the differential equation is obtained by making the following substitutions in Eq. (3.3.22):

$$\begin{cases} Y(t) \rightarrow Y_{k+1}, \\ Y_0 = Y(t_0) \rightarrow Y_k, \\ t_0 \rightarrow t_k, \\ t \rightarrow t_{k+1}. \end{cases} \tag{3.3.24}$$

As a first illustration of the procedure, consider again the decay differential equation of Eq. (3.3.5). The general solution is

$$y(t) = y_0 e^{-\lambda(t-t_0)}. \tag{3.3.25}$$

The substitutions of Eq. (3.3.24) give

$$y_{k+1} - y_k e^{-\lambda h} \tag{3.3.26}$$

which is just Eq. (3.3.7).

For our second example, consider the general logistic differential equation

$$\frac{dy}{dt} = \lambda_1 y - \lambda_2 y^2, \quad y(t_0) = y_0, \tag{3.3.27}$$

where λ_1 and λ_2 are constants. The solution to the initial value problem of Eq. (3.3.27) is given by the following expression

$$y(t) = \frac{\lambda_1 y_0}{(\lambda_1 - y_0 \lambda_2)e^{-\lambda_1(t - t_0)} + \lambda_2 y_0}. \tag{3.3.28}$$

Making the substitutions of Eq. (3.3.24) gives

$$y_{k+1} = \frac{\lambda_1 y_k}{(\lambda_1 - \lambda_2 y_k)e^{-\lambda_1 h} + \lambda_2 y_k}. \tag{3.3.29}$$

Additional algebraic manipulation gives

$$\frac{y_{k+1} - y_k}{\left(\frac{e^{\lambda_1 h} - 1}{\lambda_1}\right)} = \lambda_1 y_k - \lambda_2 y_{k+1} y_k. \tag{3.3.30}$$

Again, note that this form does not correspond to any of the discrete models constructed in the previous chapter using standard methods.

Observe, with $\lambda_2 = 0$ and $\lambda_1 \to -\lambda$, that Eq. (3.3.30) goes to the relation of Eq. (3.3.9). Also, we can obtain the exact difference scheme for the differential equation

$$\frac{dy}{dt} = -y^2 \tag{3.3.31}$$

by setting, in Eq. (3.3.30), $\lambda_1 = 0$ and $\lambda_2 = 1$. This gives the exact difference scheme

$$\frac{y_{k+1} - y_k}{h} = -y^{k+1} y_k. \tag{3.3.32}$$

The harmonic oscillator equation

$$\frac{d^2 y}{dt^2} + y = 0, \tag{3.3.33}$$

can be written as a system of two coupled, first-order differential equations

$$\frac{dy^{(1)}}{dt} = y^{(2)}, \tag{3.3.34a}$$

$$\frac{dy^{(2)}}{dt} = -y^{(1)}, \tag{3.3.34b}$$

where $y^{(1)}(t) = y(t)$. With the initial conditions

$$y_0^{(1)} = y^{(1)}(t_0), \quad y_0^{(2)} = y^{(2)}(t_0). \tag{3.3.35}$$

Equations (3.3.34) have the solutions

$$y^{(1)}(t) = \left(\frac{1}{2}\right)\left[y_0^{(1)} - iy_0^{(2)}\right]e^{i(t-t_0)}$$
$$+ \left(\frac{1}{2}\right)\left[y_0^{(1)} + iy_0^{(2)}\right]e^{-i(t-t_0)}, \tag{3.3.36}$$

$$iy^{(2)}(t) = -\left(\frac{1}{2}\right)\left[y_0^{(1)} - iy_0^{(2)}\right]e^{i(t-t_0)}$$
$$+ \left(\frac{1}{2}\right)\left[y_0^{(1)} + iy_0^{(2)}\right]e^{-i(t-t_0)}. \tag{3.3.37}$$

Making the substitutions of Eq. (3.3.24) gives

$$y_{k+1}^{(1)} = \cos(h)y_k^{(1)} + \sin(h)y_k^{(2)}, \tag{3.3.38}$$

$$y_{k+1}^{(2)} = \sin(h)y_k^{(1)} + \cos(h)y_k^{(2)}. \tag{3.3.39}$$

Finally, eliminating $y_k^{(2)}$ gives the expression

$$\frac{y_{k+1} - 2y_k + y_{k-1}}{4\sin^2\left(\frac{h}{2}\right)} + y_k = 0, \tag{3.3.40}$$

which is the exact finite difference scheme for the harmonic oscillator. Note that if $h \to \omega h$ then Eq. (3.3.40) becomes Eq. (3.3.17).

Without giving the details, we now present several other ordinary differential equations and their exact discrete models [11]:

$$2\frac{dy}{dt} + y = \frac{1}{y}, \tag{3.3.41a}$$

$$2\left[\frac{y_{k+1} - y_k}{1 - e^{-h}}\right] + \frac{y_k^2}{\left(\frac{y_{k+1}+y_k}{2}\right)} = \frac{1}{\left(\frac{y_{k+1}+y_k}{2}\right)}; \tag{3.3.41b}$$

$$\frac{dy}{dt} = -y^3, \tag{3.3.42a}$$

$$\frac{y_{k+1} - y_k}{h} = -\left[\frac{2y_{k+1}}{y_{k+1} + y_k}\right]y_{k+1}y_k^2; \tag{3.3.42b}$$

$$\frac{d^2y}{dt^2} = \lambda\frac{dy}{dt}, \tag{3.3.43a}$$

$$\frac{y_{k+1} - 2y_k + y_{k-1}}{\left(\frac{e^{\lambda h}-1}{\lambda}\right)h} = \lambda - \left(\frac{y_k - y_{k-1}}{h}\right). \tag{3.3.43b}$$

All of the above examples of exact finite difference schemes have been obtained for ordinary differential equations. We now turn to an example of a partial differential equation for which an exact discrete model exists.

Consider the nonlinear reaction-advection equation

$$u_t + u_x = u(1 - u),$$
(3.3.44)

with the initial value

$$u(x, 0) = f(x),$$
(3.3.45)

where $f(x)$ is bounded with a bounded derivative. The nonlinear transformation [1]

$$u(x, t) = \frac{1}{w(x, t)}$$
(3.3.46)

reduces Eq. (3.3.44) to the linear equation

$$w_t + w_x = 1 - w.$$
(3.3.47)

The general solution of this equation can be easily determined by standard methods [8]. It is

$$w(x, t) = g(x - t)e^{-t} + 1,$$
(3.3.48)

where $g(x)$ is an arbitrary function of z having a bounded first derivative. Imposing the initial condition of Eq. (3.3.45) allows g to be calculated, i.e.,

$$g(x) + 1 = \frac{1}{f(x)}$$
(3.3.49)

or

$$g(x) = \frac{1 - f(x)}{f(x)}.$$
(3.3.50)

Using this result with Eqs. (3.3.46) and (3.3.48), we can obtain the solution to Eqs. (3.3.44) and (3.3.45); it is given by the expression

$$u(x, t) = \frac{f(x - t)}{e^{-t} + (1 - e^{-t})f(x - t)}.$$
(3.3.51)

To proceed, we first construct the exact finite difference scheme for the unidirectional wave equation

$$u_t + u_x = 0.$$
(3.3.52)

The general solution of this equation is [8]

$$u(x, t) = H(x - t),$$
(3.3.53)

where H is an arbitrary function. Now the partial difference equation

$$u_m^{k+1} = u_{m-1}^k \qquad (3.3.54)$$

has as its general solution an arbitrary function of $(m - k)$ [10], i.e.,

$$u_m^k = F(m - k). \qquad (3.3.55)$$

If we impose the condition

$$\Delta x = \Delta t, \qquad (3.3.56)$$

then Eq. (3.3.54) can be rewritten in the following form

$$\frac{u_m^{k+1} - u_m^k}{\beta(\Delta t)} + \frac{u_m^k - u_{m-1}^k}{\beta(\Delta x)} = 0, \qquad (3.3.57)$$

where $\beta(z)$ has the property

$$\beta(z) = z + O(z^2), \quad z \to 0. \qquad (3.3.58)$$

The general solution of Eq. (3.3.57), which is formally equivalent to Eq. (3.3.54), is

$$\begin{aligned} u_m^k &= F_1[h(m - k)] \quad (h = \Delta X = \Delta t) \\ &= F_1(x_m - t_k), \end{aligned} \qquad (3.3.59)$$

where F_1 is an arbitrary function of its argument. Thus, the exact finite difference scheme for the unidirectional wave equation is Eq. (3.3.57).

We can use this result to calculate the exact difference scheme for Eq. (3.3.44). Solving Eq. (3.3.51) for $f(x - t)$ gives

$$f(x - t) = \frac{e^{-t} u(x, t)}{1 - (1 - e^{-t}) u(x, t)}. \qquad (3.3.60)$$

Now make the following substitutions in the last equation

$$\begin{cases} x \to x_m = (\Delta x)m, \quad t \to t_k = (\Delta t)k, \quad \Delta x = \Delta t = h, \\ u(x, t) \to u_m^k, \\ f(x - t) \to f[h(m - k)] = f_m^k. \end{cases} \qquad (3.3.61)$$

Doing this gives

$$f_m^k = \frac{e^{-hk} u_m^k}{1 - (1 - e^{-hk}) u_m^k}. \qquad (3.3.62)$$

However, from Eqs. (3.3.54) and (3.3.55), we know that f_m^k, satisfies the following partial difference equation

$$f_m^{k+1} = f_{m-1}^k. \qquad (3.3.63)$$

Therefore, we have

$$\frac{e^{-k(k+1)}u_m^{k+1}}{1 - [1 - e^{-h(k+1)}]u_m^{k+1}} = \frac{e^{-kk}u_{m-1}^k}{1 - (1 - e^{-hk})u_{m-1}^k}. \tag{3.3.64}$$

After some algebraic manipulations, this expression becomes

$$\frac{u_m^{k+1} - u_m^k}{e^{\Delta t} - 1} + \frac{u_m^k - u_{m-1}^k}{e^{\Delta x} - 1} = u_{m-1}^k(1 - u_m^{k+1}), \tag{3.3.65a}$$

$$\Delta t = \Delta x. \tag{3.3.65b}$$

Discrete models of the nonlinear reaction-advection equation using the standard rules do not have the structure of Eqs. (3.3.65). For example, a particular standard model is

$$\frac{u_m^{k+1} - u_m^k}{\Delta t} + \frac{u_{m+1}^k - u_m^k}{\Delta x} = u_m^k(1 - u_m^k). \tag{3.3.66}$$

3.4   Nonstandard Modeling Rules

Let us now examine in detail the results obtained in the previous section. In particular, we concentrate on the exact finite difference schemes for the general logistic ordinary differential equation and the nonlinear, reaction-advection partial differential equation. These are given, respectively, by Eqs. (3.3.27) and (3.3.30), and (3.3.44) and (3.3.65). The following observations are important:

(i) Exact finite difference schemes generally require that nonlinear terms be modeled nonlocally. Thus, for the logistic equation the y^2 term is evaluated at two different grid points

$$y^2 \to y_{k+1}y_k. \tag{3.4.1}$$

Similarly, the u^2 term for the nonlinear, reaction-advection equation is modeled by the expression

$$u^2 \to u_{m-1}^k u_m^{k+1}. \tag{3.4.2}$$

This corresponds to u being evaluated at two different lattice space-points and two different lattice time-points. Note that

$$\underset{\substack{h \to 0 \\ k \to \infty \\ hk=t=\text{fixed}}}{\text{Lim}} y_{k+1}y_k = \underset{\substack{h \to 0 \\ k \to \infty \\ hk=t=\text{fixed}}}{\text{Lim}} y_k^2 = y(t)^2, \tag{3.4.3}$$

and

$$\lim_{\substack{\Delta x \to 0 \\ \Delta t \to 0 \\ k \to \infty \\ m \to \infty \\ (\Delta x)m=x=\text{fixed} \\ (\Delta t)k=t=\text{fixed}}} u_{m-1}^k u_m^{k+1} = \lim_{\substack{\Delta x \to 0 \\ \Delta t \to 0 \\ k \to \infty \\ m \to \infty \\ (\Delta x)m=x=\text{fixed} \\ (\Delta t)k=t=\text{fixed}}} (u_m^k)^2 = [u(x,t)]^2. \qquad (3.4.4)$$

However, for finite, fixed, nonzero values of the step-sizes, the two representations of the squared terms in Eqs. (3.4.3) and (3.4.4) are not equal, i.e.,

$$y_{k+1}y_k \neq (y_k)^2, \qquad (3.4.5a)$$

$$u_{m-1}^k u_m^{k+1} \neq (u_m^k)^2. \qquad (3.4.5b)$$

Therefore, a seemingly trivial modification in the modeling of nonlinear terms can lead to major changes in the solution behaviors of the difference equations.

(ii) The discrete derivatives for both differential equations have denominator functions that are more complicated than those used in the standard modeling procedure. For example, the time-derivative in the logistic equation is replaced by the following discrete representation

$$\frac{dy}{dt} \to \frac{y_{k+1} - y_k}{\left(\frac{e^{\lambda_1 h} - 1}{\lambda_1}\right)}. \qquad (3.4.6)$$

Thus, the denominator function depends on both the parameter λ_1 and the step-size $h = \Delta t$.

(iii) In the discrete modeling of partial differential equations, functional relations may exist between the various step-sizes. For the nonlinear, reaction-advection equation, the required restriction is $\Delta t = \Delta x$.

(iv) Of importance is the observation that for partial differential equations, the modeling of first-derivatives may require the use of a forward-Euler type discrete derivative for the time variable, but, a backward-Euler type discrete derivative for the space variable. See Eq. (3.3.65).

(v) Finally, we found that the order of the discrete derivatives in exact finite difference schemes is always equal to the corresponding order of the derivatives of the differential equation.

With the above facts in hand, we now study the various sources of numerical instabilities for standard models of the logistic and nonlinear, reaction-advection equations. First, consider the following finite difference scheme for the logistic equation

$$\frac{y_{k+1} - y_k}{h} = y_k(1 - y_k). \qquad (3.4.7)$$

This discrete representation is expected to have numerical instabilities for two reasons: (a) the denominator function is incorrect; (b) the nonlinear term is modeled locally on the grid. See Eq. (3.3.30) for comparison with the exact scheme. Now consider the following model for the nonlinear, reaction-advection equation

$$\frac{u_m^{k+1} - u_m^k}{\Delta t} + \frac{u_{m+1}^k - u_{m-1}^k}{2\Delta x} = u_m^k(1 - u_m^k). \qquad (3.4.8)$$

There are several sources of numerical in stabilities: (a) The nonlinear term is modeled locally on the computational grid. (b) The first-order space derivative is modeled by a higher order central difference scheme. (c) There is no explicit relation indicated between the space and time step-sizes. Again, comparison to Eq. (3.3.65) should be made.

These results can be used to understand the findings of Mitchell and Bruch [12] who consider the one-dimensional, nonlinear, reaction-diffusion equation

$$u_t = Du_{xx} + \alpha u(1 - u), \qquad (3.4.9)$$

where D and α are non-negative constants. This equation is known as the Fisher equation [13]. In our notation, they numerically investigated the properties of the solutions to the finite difference scheme

$$\frac{u_m^{k+1} - u_m^k}{\Delta t} = D\left[\frac{u_{m+1}^k - 2u_m^k + u_{m-1}^k}{(\Delta x)^2}\right] + \alpha u_m^k(1 - u_m^k). \qquad (3.4.10)$$

They found numerical solutions that were chaotic as well as other solutions that diverged. From the perspective of our analysis, it should be clear that these numerical instabilities were a primary consequence of the local modeling for the u^2 term. Note that in Eq. (3.4.10), the discrete space independent difference equation is the logistic difference equation.

Based on both analytical and numerical studies of exact finite difference schemes for a large number of ordinary and partial differential equations [1, 2, 3, 11], we present the following rules for the construction of discrete models.

Rule 1. The orders of the discrete derivatives must be exactly equal to the orders of the corresponding derivatives of the differential equations.

Rule 2. Denominator functions for the discrete derivatives must, in general, be expressed in terms of more complicated functions of the step-sizes than those conventionally used.

Rule 3. Nonlinear terms must, in general, be modeled nonlocally on the computational grid or lattice.

Rule 4. Special solutions of the differential equations should also be special (discrete) solutions of the finite difference models.

Rule 5. The finite difference equations should not have solutions that do not correspond exactly to solutions of the differential equations.

A major advantage of having an exact difference equation model for a differential equation is that questions related to the usual considerations of consistency, stability and convergence [9, 14, 15, 16] need not arise. However, it is essentially impossible to construct an exact discrete model for an arbitrary differential equation. This is because to do so would be tantamount to knowing the general solution of the original differential equation. However, the situation is not hopeless. The above five modeling rules can be applied to the construction of finite difference schemes. While these discrete models, in general, will not be exact schemes, they will possess certain very desirable properties. In particular, we may hope to eliminate a number of the problems related to numerical instabilities.

The next section introduces the notion of a *best finite difference scheme*. After discussion of this concept, we present the construction of best discrete models for two nonlinear differential equations.

3.5   Best Finite Difference Schemes

A *best finite difference scheme* is a discrete model of a differential equation that is constructed according to the five rules given in Section 3.4. In general, best schemes are not exact schemes. However, they offer the prospect of obtaining finite difference models that do not possess the standard numerical instabilities. As will be demonstrated in two examples, the application of the five nonstandard modeling rules does not necessarily lead to a unique discrete model for a given differential equation. We are currently studying how to resolve this difficulty.

An equation of fundamental importance in the study of one-dimensional, nonlinear oscillatory phenomena is the Duffing equation [17]

$$\frac{d^2y}{dt^2} + a\frac{dy}{dt} + by + cy^3 = F\cos\omega t, \tag{3.5.1}$$

where (a, b, c, F, ω) are constants. For our purposes, the following special case will be studied [18]

$$\frac{d^2y}{dt^2} + \omega^2 y + \lambda y^3 = 0, \tag{3.5.2}$$

where ω is the angular frequency of the linear oscillation and λ is a measure of the strength of the nonlinear term. A first-integral or energy relation is

$$\left(\frac{1}{2}\right)\left(\frac{dy}{dt}\right)^2 + \frac{\omega^2 y^2}{2} + \frac{\lambda y^4}{4} = E, \qquad (3.5.3)$$

where E is the energy constant. If λ is restricted to be non-negative, i.e.,

$$\lambda \geq 0, \qquad (3.5.4)$$

then it follows from Eq. (3.5.3) that all the solutions of Eq. (3.5.2) are bounded and periodic [17].

A standard finite difference model for Eq. (3.5.2) is

$$\frac{y_{k+1} - 2y_k + y_{k-1}}{h^2} + \omega^2 y_k + \lambda u_k^3 = 0, \qquad (3.5.5)$$

where $h = \Delta t$. This equation can be rewritten as

$$\frac{(y_{k+1} - y_k) - (y_k - y_{k-1})}{h^2} = -\omega^2 y_k - \lambda y_k^3. \qquad (3.5.6)$$

Multiplying by

$$(y_{k+1} - y_k) + (y_k - y_{k-1}) = y_{k+1} - y_{k-1} \qquad (3.5.7)$$

gives

$$\frac{(y_{k+1} - y_k)^2}{h^2} - \frac{(y_k - y_{k-1})^2}{h^2} = -\omega^2 (y_{k+1} y_k - y_k y_{k-1})$$
$$- \lambda(y_{k+1} y_k^3 - y_{k-1} y_k^3). \qquad (3.5.8)$$

The transposition of certain terms gives the following expression

$$\frac{(y_{k+1} - y_k)^2}{h^2} + \omega^2 y_{k+1} y_k + \lambda y_{k+1} y_k^3$$
$$= \frac{(y_k - y_{k-1})^2}{h^2} + \omega^2 y_k y_{k-1} + \lambda y_{k-1} y_k^3. \qquad (3.5.9)$$

If energy is to be conserved, then the right-side of this equation should reduce to the terms on the left-side when k is replaced by $k + 1$. The first two terms do have this property; however, the third term on the right-side does not become the third term on the left-side under this transformation. Therefore, we conclude that the standard finite difference scheme of Eq. (3.5.5) does not conserve energy [18].

The application of the five rules from the previous section gives the following discrete model for Eq. (3.5.2):

$$\frac{y_{k+1} - 2y_k + y_{k-1}}{\psi} + \omega^2 y_k + \lambda u_k^2 \left(\frac{y_{k+1} + y_{k-1}}{2}\right) = 0, \qquad (3.5.10)$$

where ψ has the property that

$$\psi(h,\omega,\lambda) = h^2 + O(h^4), \quad h \to 0. \tag{3.5.11}$$

Note that in the limits

$$h \to 0, \quad k \to \infty, \quad hk = t = \text{constant}, \tag{3.5.12}$$

this finite difference equation converges to the original Duffing equation as given by Eq. (3.5.2). Also, observe that the nonlinear term y^3 in the differential equation has the following discrete representation

$$y^3 \to y_k^2 \left(\frac{y_{k+1} + y_{k-1}}{2} \right). \tag{3.5.13}$$

To proceed further, it must now be shown that the discrete model of Eq. (3.5.10) satisfies a conservation law. This is easily done by following the same steps as presented above for Eq. (3.5.5). We find the result

$$\frac{(y_{k+1} - y_k)^2}{2\psi} + \left(\frac{1}{2} \right) \omega^2 y_{k+1} y_k + \left(\frac{1}{4} \right) \lambda y_{k+1}^2 y_k^2$$
$$= \frac{(y_k - y_{k-1})^2}{2\psi} + \left(\frac{1}{2} \right) \omega^2 y_k y_{k-1} + \left(\frac{1}{4} \right) \lambda y_k^2 y_{k-1}^2. \tag{3.5.14}$$

Observe that the transformation $k \to k + 1$ changes the right-side of Eq. (3.5.14) into the expression on the left-side. This means that, independent of the value of k, each side of Eq. (3.5.14) is equal to the same constant. Consequently, the discrete model of the Duffing equation, given by Eq. (3.5.10), has the following associated conservation law

$$\left(\frac{1}{2} \right) \frac{(y_{k+1} - y_k)^2}{\psi} + \left(\frac{1}{2} \right) \omega^2 y_{k+1} y_k + \left(\frac{\lambda}{4} \right) y_{k+1}^2 y_k^2 = \text{constant}. \tag{3.5.15}$$

This is to be compared to the energy relation of the differential equation as expressed by Eq. (3.5.3).

An ambiguity in the above modeling process is that the denominator function ψ is not uniquely specified. At this level of analysis, any function that obeys the relation given by Eq. (3.5.11) works. Finally, it should be indicated that Potts [19] has also investigated various nonstandard finite difference approximations to the unforced, undamped Duffing differential equation.

For our second example of the construction of a best finite difference scheme, we consider the nonlinear diffusion equation [11]

$$u_t = u u_{xx}. \tag{3.5.16}$$

No known exact solution exists for the general initial-value problem for this equation. However, a special rational solution is known. To obtain it, write $u(x,t)$ in the form

$$u(x,t) = C(t)D(x). \qquad (3.5.17)$$

Substitution into Eq. (3.5.16) gives

$$C'D = CDCD'' \qquad (3.5.18)$$

and

$$\frac{C'(t)}{C^2} = D''(x) = \alpha, \qquad (3.5.19)$$

where α is the separation constant. Integrating these differential equations gives the solutions

$$C(t) = \frac{1}{\alpha_1 - \alpha t}, \qquad (3.5.20)$$

$$D(x) = \left(\frac{\alpha}{2}\right) x^2 + \beta_1 x + \beta_2, \qquad (3.5.21)$$

where $(\alpha_1,\ \beta_1,\ \beta_2)$ are arbitrary integration constants. Therefore, Eq. (3.5.16) has the following special rational solution

$$u(x,t) = \frac{\left(\frac{\alpha}{2}\right) x^2 + \beta_1 x + \beta_2}{\alpha_1 - \alpha t}. \qquad (3.5.22)$$

A nonstandard explicit finite difference model for Eq. (3.5.16) is

$$\frac{u_m^{k+1} - u_m^k}{\Delta t} = u_m^{k+1}\left[\frac{u_{m+1}^k - 2u_m^k + u_{m-1}^k}{(\Delta x)^2}\right], \qquad (3.5.23)$$

where the orders of the discrete derivatives are the same as the derivatives of the partial differential equation. Also, the nonlinear term, uu_{xx} is modeled nonlocally on the lattice, i.e., one term is at time-step k and the other is at $k+1$.

Since we know a special solution of the original partial differential equation, we must require that any best finite difference scheme also have this as a solution. To see whether this is the case for the model of Eq. (3.5.23), let us calculate a special solution by assuming that u_m^k has the form

$$u_m^k = C^k D_m. \qquad (3.5.24)$$

Substitution of Eq. (3.5.24) into Eq. (3.5.23) gives

$$\frac{(C^{k+1} - C^k)D_m}{\Delta t} = C^{k+1}C^k D_m\left[\frac{D_{m+1} - 2D_m + D_{m-1}}{(\Delta x)^2}\right], \qquad (3.5.25)$$

and

$$\frac{(C^{k+1} - C^k)D_m}{(\Delta t)C^{k+1}C^k} = \frac{D_{m+1} - 2D_m + D_{m-1}}{(\Delta x)^2} = \alpha, \qquad (3.5.26)$$

where α is the separation constant. The two difference equations

$$C^{k+1} - C^k = \alpha(\Delta t)C^{k+1}C^k, \qquad (3.5.27)$$

$$D_{m+1} - 2D_m + D_{m-1} = \alpha(\Delta)^2, \qquad (3.5.28)$$

have solutions that can be put in the forms [10]

$$C^k = \frac{1}{\alpha_1 - \alpha t_k}, \qquad (3.5.29)$$

$$D_m = \left(\frac{\alpha}{2}\right) x_m^2 + \beta x_m + \beta_2, \qquad (3.5.30)$$

where

$$t_k = (\Delta t)k, \quad x_m = (\Delta x)m, \qquad (3.5.31)$$

and $(\alpha_1, \beta_1, \beta_2)$ are arbitrary constants. Thus, a special solution to the partial difference equations given in Eq. (3.5.23) is

$$u_m^k = \frac{\left(\frac{\alpha}{2}\right) x_m^2 + \beta_1 x_m + \beta_2}{\alpha_1 - \alpha t_k}. \qquad (3.5.32)$$

For these special solutions

$$u(x_m, t_k) = u_m^k. \qquad (3.5.33)$$

Thus, we conclude that Eq. (3.5.23) is a best finite difference scheme for Eq. (3.5.16).

It should be acknowledged, however, that another best finite difference scheme also exists for Eq. (3.5.16) and is given by the implicit scheme

$$\frac{u_m^{k+1} - u_m^k}{\Delta t} + u_m^k \left[\frac{u_{m+1}^{k+1} - 2u_m^{k+1} + u_{m-1}^{k+1}}{(\Delta x)^2}\right] = 0. \qquad (3.5.34)$$

By the method of separation-of-variables, it is easily determined that this partial difference equation has the expression of Eq. (3.5.32) as a solution.

At this stage of the discrete modeling process, there is no reason to choose one form of best finite difference scheme over the other. This result illustrates the (sometimes) ambiguities that arise even in the application of the nonstandard modeling rules. Additional information on the properties of the solutions to the partial differential equation is needed to make a (possible) unique selection.

Throughout the remainder of this book, we will use the five nonstandard modeling rules and possible modifications of them (when needed) to construct discrete models for a variety of special classes of differential equations.

References

1. R. E. Mickens, *Numerical Methods for Partial Differential Equations* **5**, 313–325 (1989). Exact solutions to a finite-difference model of a nonlinear reaction-advection equation: Implications for numerical analysis.
2. R. E. Mickens and A. Smith, *Journal of the Franklin Institute* **327**, 143–149 (1990). Finite-difference models of ordinary differential equations: Influence of denominator functions.
3. J. Marsden and A. Weinsten, *Calculus I* (Springer-Verlag, New York, 2nd edition, 1985), section 1.3.
4. R. B. Potts, *American Mathematical Monthly* **89**, 402–407 (1982). Differential and difference equations.
5. V. V. Nemytski and V. V. Stepanov, *Qualitative Theory of Differential Equations* (Princeton University Press; Princeton, NJ; 1969).
6. J. A. Murdock, *Perturbations: Theory and Methods* (Wiley-Interscience, New York, 1991), Appendix D.
7. R. E. Mickens, Difference equation models of differential equations having zero local truncation errors, in *Differential Equations*, I. W. Knowles and R. T. Lewis, editors (North-Holland, Amsterdam, 1984), pp. 445–449.
8. E. Zauderer, *Partial Differential Equations of Applied Mathematics* (Wiley-Interscience, New York, 1983).
9. F. B. Hilderbrand, *Finite-Difference Equations and Simulations* (Prentice-Hall; Englewood Cliffs, NJ; 1968).
10. R. E. Mickens, *Difference Equations: Theory and Applications* (Van Nostrand Reinhold, New York, 1990), section 4.3.
11. R. E. Mickens, *Mathematics and Computer Modelling* **11**, 528–530 (1988). Different equation models of differential equations.
12. A. R. Mitchell and J. C. Bruch, *Numerical Methods for Partial Differential Equations* **1**, 13–23 (1985). A numerical study of chaos in a reaction-diffusion equation.
13. J. Murray, *Lectures on Nonlinear Differential Equation Models in Biology* (Clarendon Press, Oxford, 1977).
14. R. D. Richtmyer and K. W. Morton, *Difference Methods for Initial-Value Problems* (Wiley-Interscience, New York, 2nd edition, 1967).
15. A. R. Mitchell and D. F. Griffiths, *Finite Difference Methods in Partial Differential Equations* (Wiley, New York, 1980).

16. M. K. Jain, *Numerical Solution of Differential Equations* (Halsted Press, New York, 1984).

17. J. J. Stokes, *Nonlinear Vibrations* (Interscience, New York, 1950).

18. R. E. Mickens, O. Oyedeji and C. R. McIntyre, *Journal of Sound and Vibration* **130**, 509–512 (1989). A difference-equation model of the Duffing equation.

19. R. B. Potts, *Journal of the Australian Mathematics Society (Series B)* **23**, 349–356 (1982). Best difference equation approximation to Duffing's equation.

Chapter 4

First-Order ODE's

4.1 Introduction

This chapter presents a technique for constructing finite difference models of a single scalar differential equation. The work to be discussed is based on the investigations of Mickens and Smith [1] and Mickens [2]. The equation to be investigated is the autonomous first-order differential equation

$$\frac{dy}{dt} = f(y). \tag{4.1.1}$$

Our analysis will be done under the assumption that

$$f(y) = 0, \tag{4.1.2}$$

has only simple zeros. Our goal is to construct discrete models of Eq. (4.1.1) that do not exhibit elementary numerical instabilities, i.e., solutions to the finite difference equation that do not correspond to any of the solutions to the differential equation. For Eq. (4.1.1) numerical instabilities will occur whenever the linear stability properties of any of the fixed-points for the difference scheme differs from that of the differential equation [3, 4, 5].

The main purpose of this chapter is to prove, for Eq. (4.1.1), that it is possible to construct finite difference schemes that have the correct linear stability properties for all finite step-sizes [1, 2]. The proof involves the explicit construction of such schemes.

This chapter extends the analysis given in the latter half of Section 2.7. In summary, we found there that numerical instabilities occur by several mechanisms: (a) For a central difference scheme, the numerical instabilities are a consequence of the order of the difference scheme being higher than the order of the differential equation. (b) For forward-Euler schemes, the numerical instabilities arise when the step-size is larger than some fixed, finite value, $h > h^* > 0$. (c) The implicit backward-Euler scheme exhibits super-stability, i.e., its numerical instabilities occur above some threshold value of the step-size, say $h > h_0$, such that all the fixed-points of the difference scheme become stable. (d) The use of higher order (in h) schemes, such as Runge-Kutta methods, gives rise to numerical instabilities because of the appearance of spurious real fixed-points for $h > h_1$.

4.2 A New Finite Difference Scheme

Denote the fixed-points of Eq. (4.1.1) by

$$\{\bar{y}^{(i)};\ i = 1, 2, \dots, I\}, \tag{4.2.1}$$

where I may be unbounded. The fixed-points are the solutions to the equation

$$f(y) = 0. \tag{4.2.2}$$

Define R_i to be

$$R_i \equiv \frac{df[\bar{y}^{(i)}]}{dy}, \tag{4.2.3}$$

and R^* as

$$R^* \equiv \text{Max}\{|R_i|;\ i = 1, 2, \dots, I\}. \tag{4.2.4}$$

Linear stability analysis applied to the i-th fixed-point gives the following results [6]:

  (i) If $R_i > 0$, the fixed-point $y(t) = \bar{y}^{(i)}$ is linearly unstable.
  (ii) If $R_i < 0$, the fixed-point $y(t) = \bar{y}^{(i)}$ is linearly stable.

Now consider the following finite difference scheme for Eq. (4.1.1)

$$\frac{y_{k+1} - y_k}{\left[\frac{\phi(hR^*)}{R^*}\right]} = f(y_k), \tag{4.2.5}$$

where $\phi(z)$ has the two properties

$$\phi(z) = z + O(z^2), \quad z \to 0, \tag{4.2.6a}$$
$$0 < \phi(z) < 1, \quad z > 0. \tag{4.2.6b}$$

Theorem. *The finite difference scheme of Eq. (4.2.5) has fixed-points with exactly the same linear stability properties as the differential equation*

$$\frac{dy}{dt} = f(y), \tag{4.2.7}$$

for all $h > 0$.

Proof. Represent a perturbation about the i-th fixed-point by

$$y_k = \bar{y}^{(i)} + \epsilon_k. \tag{4.2.8}$$

The linear stability analysis equation for ϵ_k is

$$\frac{\epsilon_{k+1} - \epsilon_k}{\left[\frac{\phi(hR^*)}{R*}\right]} = R_i \epsilon_k \tag{4.2.9}$$

or

$$\epsilon_{k+1} = \left[1 + \left(\frac{R_i}{R^*}\right) \phi(hR^*)\right] \epsilon_k, \tag{4.2.10}$$

which has the solution

$$\epsilon_k = \epsilon_0 \left[1 + \left(\frac{R_i}{R^*}\right) \phi(hR^*)\right]^k. \tag{4.2.11}$$

For $R_i > 0$, the fixed-point of the differential equation is linearly unstable. Thus, it follows that

$$1 + \left(\frac{R_i}{R^*}\right) \phi(hR^*) > 0, \quad h > 0. \tag{4.2.12}$$

Therefore, $y_k = \bar{y}^{(i)}$ is also linearly unstable for $h > 0$.

For $R_i < 0$, the fixed-point of the differential equation is linearly stable. In this instance

$$0 < 1 - \left(\frac{|R_i|}{R^*}\right) \phi(hR) < 1, \quad h > 0, \tag{4.2.13}$$

a result that follows directly from Eqs. (4.2.4) and (4.2.6). Therefore, $y_k = \bar{y}^{(i)}$ is linearly stable for $h > 0$.

This theorem shows that it is possible to construct discrete models for a single scalar, autonomous ordinary differential equation such that no elementary numerical instabilities occur in their solutions. This result is related to the fact that most elementary numerical instabilities arise from a given fixed-point having the opposite linear stability properties in the difference scheme and the differential equation. The above construction demonstrates that to achieve the correct linear stability behavior, a generalized definition of the discrete derivative must be used [1]. None of the standard finite difference modeling procedures have this property, namely, the correct linear stability behavior for all step-sizes.

The above finite difference scheme uses the following *denominator function* for the discrete first-derivative

$$D(h, R^*) \equiv \frac{\phi(hR^*)}{R^*}, \tag{4.2.14}$$

where ϕ and R^* are given by Eqs. (4.2.4) and (4.2.6). This form replaces the simple "h" function found in the standard finite difference schemes, i.e.,

$$\frac{dy}{dt} \to \frac{y_{k+1} - y_k}{h}, \quad \text{(standard schemes).} \qquad (4.2.15)$$

Note that in the limits ($h \to 0$, $k \to \infty$, $hk = t = $ fixed), the generalized discrete derivative reduces to the first derivative, i.e.,

$$\operatorname*{Lim}_{\substack{h \to 0 \\ k \to \infty \\ hk=t=\text{fixed}}} \frac{y_{k+1} - y_k}{\left[\frac{\phi(hR^*)}{R^*}\right]} = \frac{dy}{dt}. \qquad (4.2.16)$$

However, for fixed $h > 0$ and a given value of k, the generalized discrete derivative may have a numerical magnitude that differs greatly from the standard discrete derivatives such as those given by the central difference, forward-Euler and backward-Euler representations.

The denominator function $D(h, R^*)$ can be considered a renormalization of the step-size h to a new value h', i. e.,

$$h \to h' = D(h, R^*). \qquad (4.2.17)$$

This concept of renormalized variables and constants occurs frequently in various areas of the sciences [6].

It should be noted that, except for the requirements given in Eqs. (4.2.6), the function $\phi(z)$ can be arbitrary. However, a particularly useful and simple functional form for $\phi(z)$ is the following expression which occurs in many exact finite difference schemes

$$\phi(z) = 1 - \epsilon^{-z}. \qquad (4.2.18)$$

Finally, it should be observed that the difference scheme of Eq. (4.2.5) resolves the problem of "super-stability" that occurs in the use of the backward-Euler scheme [3, 4, 7, 8, 9]; see Section 2.7. The finite difference scheme, for this case, is

$$\frac{y_{k+1} - y_k}{\left[\frac{\phi(hR^*)}{R^*}\right]} = f(y_{k+1}). \qquad (4.2.19)$$

For a perturbation about the i-th fixed-point

$$y_k = \bar{y}^{(i)} + \epsilon_k, \qquad (4.2.20)$$

the equation for ϵ_k is

$$\frac{\epsilon_{k+1} - \epsilon_k}{\left[\frac{\phi(hR^*)}{R^*}\right]} = R_i \epsilon_{k+1}, \qquad (4.2.21)$$

which has the solution

$$\epsilon_k = \epsilon_0 \left[\frac{1}{1 - \left(\frac{R(i}{R^*}\right) \phi(hR^*)} \right]^k . \tag{4.2.22}$$

If $R_i < 0$, the fixed-point for the differential equation is linearly stable. Now, for this case

$$1 - \left(\frac{R(i}{R^*}\right) \phi(hR^*) = 1 + \left(\frac{|R(i|}{R^*}\right) \phi(hR^*) > 0. \tag{4.2.23}$$

This implies that $y_k = \bar{y}^{(i)}$ is also linearly stable for $h > 0$. Likewise, for $R_i > 0$, the fixed-point for the differential equation is linearly unstable. Since

$$0 < \frac{R_i}{R^*} < 1; \quad 0 < \phi < 1, \quad 0 < h < \infty; \tag{4.2.24}$$

it follows that

$$0 < 1 - \left(\frac{R_i}{R^*}\right) \phi(hR^*) < 1, \tag{4.2.25}$$

and it can be concluded that $y_k = \bar{y}^{(i)}$ is linearly unstable. Thus, the finite difference scheme of Eq. (4.2.19) has fixed-points with exactly the same linear stability properties as the scalar differential equation. This result holds for all step-sizes, $h > 0$. Consequently, "super-stability" will not occur in the discrete model of Eq. (4.2.19).

4.3 Examples

We illustrate the power of the new finite difference scheme for scalar first-order ordinary differential equations by applying it to three equations: the decay equation, the logistic differential equation and an equation having three fixed-points.

4.3.1 *Decay Equation*

The decay differential equation is

$$\frac{dy}{dt} = -\lambda y, \tag{4.3.1}$$

where λ is a positive constant. The function $f(y)$ is

$$f(y) = -\lambda y. \tag{4.3.2}$$

There is a single globally stable fixed-point at $\bar{y}^{(i)} = 0$. In addition,

$$R_1 = -\lambda, \quad R^* = \lambda. \tag{4.3.3}$$

Now, select for $\phi(z)$, the expression

$$\phi(z) = 1 - e^{-z}. \tag{4.3.4}$$

Substitution of these results into Eq. (4.2.5) gives

$$\frac{y_{k+1} - y_k}{\left(\frac{1 - e^{-\lambda h}}{\lambda}\right)} = -\lambda y_k, \tag{4.3.5}$$

which is the exact finite difference scheme for the decay equation; see Eq. (3.3.9).

Thus, in this case, the new finite difference scheme gives the exact discrete model.

4.3.2   *Logistic Equation*

For the logistic differential equation

$$\frac{dy}{dt} = y(1 - y), \tag{4.3.6}$$

we have

$$f(y) = y(1 - y), \tag{4.3.7}$$

two fixed-points

$$\bar{y}^{(1)} = 0, \quad \bar{y}^{(2)} = 1, \tag{4.3.8}$$

and

$$R_1 = 1, \quad R_2 = -1, \quad R^* = 1. \tag{4.3.9}$$

The substitution of Eqs. (4.3.7), (4.3.9) and (4.3.4) into Eq. (4.2.5) gives

$$\frac{y_{k+1} - y_k}{1 - e^{-h}} = -y_k(1 - y_k). \tag{4.3.10}$$

Figure 4.3.1 gives the numerical solution of the logistic differential equation using Eq. (4.3.10). The initial condition is $y(0) = y_0 = 0.5$ and the step-sizes used in the computation are $h = (0.01, 1.5, 2.5)$. Note that for all three step-sizes, the numerical solution has the same qualitative behavior as the exact solution, i.e., a monotonic increase to the value one.

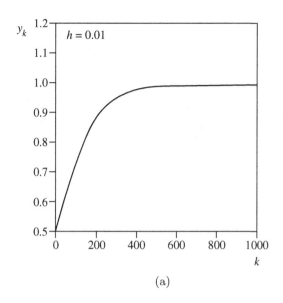

(a)

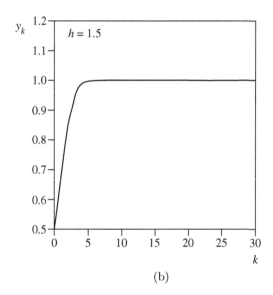

(b)

Fig. 4.3.1   Plots of Eq. (4.3.10). For each graph $y = 0.5$, (a) $h = 0.01$, (b) $h = 1.5$.

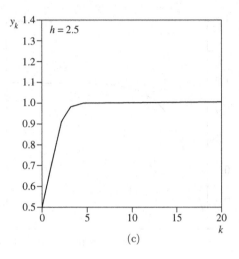

Fig. 4.3.1   Plots of Eq. (4.3.10). For each graph $y = 0.5$, (c) $h = 2.5$.

4.3.3   *ODE with Three Fixed-Points*

The simplest ordinary differential equation with three fixed-points is

$$\frac{dy}{dt} = y(1 - y^2). \tag{4.3.11}$$

For this equation

$$f(y) = y(1 - y^2), \tag{4.3.12}$$

$$\bar{y}^{(1)} = 0, \quad \bar{y}^{(2)} = 1, \quad \bar{y}^{(3)} = -1, \tag{4.3.13}$$

$$R_1 = 1, \quad R_2 = R_3 = -2, \quad R^* = 2. \tag{4.3.14}$$

Using $\phi(z)$ from Eq. (4.3.4), we obtain, on substitution of these results into Eq. (4.2.5), the following discrete model for Eq. (4.3.11)

$$\frac{y_{k+1} - y_k}{\left(\frac{1 - e^{-2h}}{2}\right)} = y_k(1 - y_k^2). \tag{4.3.15}$$

Figure 4.3.2 gives the general behavior of the solutions for various initial values, $y(0) = y_0$. The $(\pm)$ sign denotes the regions where the derivative has a constant sign; at the fixed-point, the derivative must be zero. For $y_0 > 0$, all solutions approach the stable fixed-point at $\bar{y}^{(2)} = 1$. Likewise, for $y_0 < 0$, all solutions approach the other stable fixed-point at $\bar{y}^{(3)} = -1$.

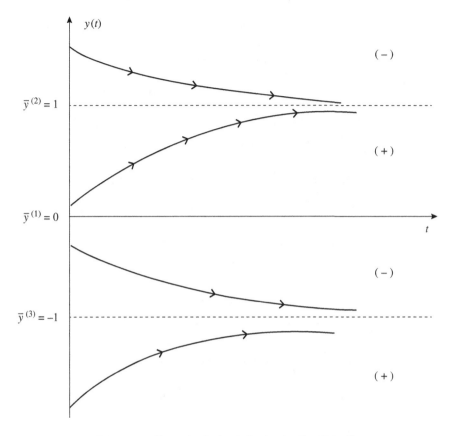

Fig. 4.3.2   General solution behavior for Eq. (4.3.11).

Figure 4.3.3 presents numerical solutions obtained from Eq. (4.3.15). Each graph starts with the initial condition $y(0) = y_0 = 0.5$. The four step-sizes used are $h = (0.01, 0.75, 1.5, 2.5)$. Observe that for all four step-sizes, the numerical functions have exactly the same qualitative behavior as the corresponding solution of the differential equation, i.e., a monotonic increase from $y_0 = 0.5$ to $y_\infty = \bar{y}^{(2)} = 1$.

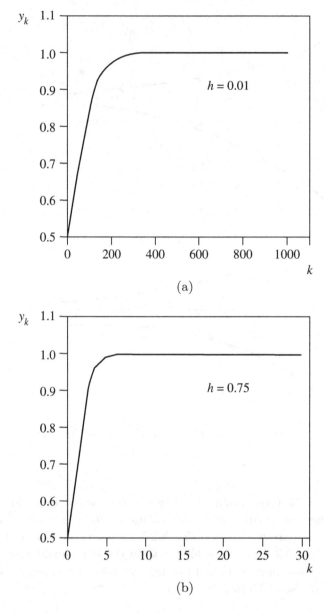

Fig. 4.3.3   Plots of Eq. (4.3.15). For each graph $y_0 = 0.5$, (a) $h = 0.01$. (b) $h = 0.75$.

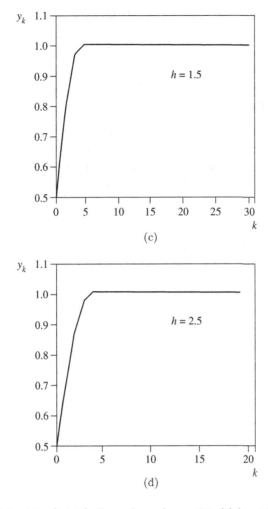

Fig. 4.3.3 Plots of Eq. (4.3.15). For each graph $y_0 = 0.5$, (c) $h = 1.5$. (d) $h = 2.5$.

For purposes of comparison, we consider the standard forward-Euler scheme for Eq. (4.3.11); it is

$$\frac{y_{k+1} - y_k}{h} = y_k(1 - y_k^2). \qquad (4.3.16)$$

Perturbations about the three fixed-points

$$y_k = \begin{cases} 1 + \eta_k \\ 0 + \epsilon_k \\ -1 + \eta_k, \end{cases} \qquad (4.3.17)$$

give the following linear stability equations

$$\epsilon_{k+1} = (1 + h)\epsilon_k, \qquad (4.3.18)$$

$$\eta_{k+1} = (1 - 2h)\eta_k. \qquad (4.3.19)$$

From Eq. (4.3.18), it follows that $\bar{y}^{(1)} = 0$ is linearly unstable for all $h > 0$. However, the fixed-points at $\bar{y}^{(2)} = 1$, $\bar{y}^{(3)} = -1$, have the following linear stability properties:

  (i) For $0 < h < 0.5$, both fixed-points are linearly stable.
 (ii) For $0.5 < h < 1$, both fixed-points are linearly stable; however, the perturbations decrease to zero with an oscillating amplitude.
(iii) For $h > 1$, the two fixed-points are linearly unstable.

The results given in Figure 4.3.4 are numerical solutions obtained from the forward-Euler scheme of Eq. (4.3.16). For each, the initial condition is $y(0) = y_0 = 0.5$ and the respective step-sizes are $h = (0.01, 0.75, 1.5, 2.0)$. Note that the graphs are fully consistent with the above linear stability analysis.

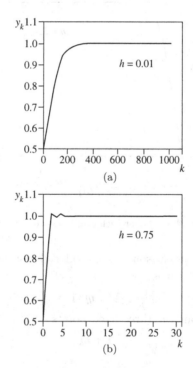

Fig. 4.3.4   Plots of Eq. (4.3.16). For each graph $y_0 = 0.5$, (a) $h = 0.01$, (b) $h = 0.75$.

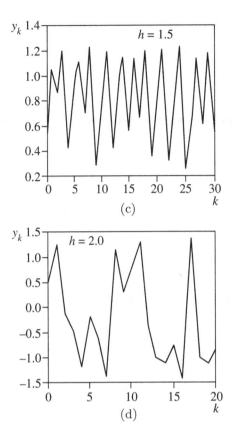

Fig. 4.3.4   Plots of Eq. (4.3.16). For each graph $y_0 = 0.5$, (c) $h = 1.5$, (d) $h = 2.0$.

The results of this section can be summarized in the statement that the new finite difference scheme of Eq. (4.2.5) provides superior discrete models of the three differential equations studied as compared to the use of the standard forward-Euler scheme.

4.4   Nonstandard Schemes

Chapter 3 provided a set of nonstandard modeling rules. We now apply them to two of the differential equations examined in the last section. The new modeling rule, to be added to the results of Section 4.2, is the requirement that nonlinear terms be modeled nonlocally on the computational grid.

4.4.1 Logistic Equation

The discrete scheme for the logistic differential equation, with a nonlocal nonlinear term, is

$$\frac{y_{k+1} - y_k}{1 - e^{-h}} = y_k(1 - y_{k+1}).\qquad(4.4.1)$$

This difference equation can be solved exactly by use of the transformation [10]

$$y_k = \frac{1}{w_k}.\qquad(4.4.2)$$

This gives

$$w_{k+1} - \left(\frac{1}{2 - e^{-h}}\right)w_k = \frac{1 - e^{-h}}{2 - e^{-h}},\qquad(4.4.3)$$

whose exact solution is [10]

$$w_k = 1 + A(2 - e^{-h})^{-k},\qquad(4.4.4)$$

where A is an arbitrary constant. Imposing the initial condition, $y(0) = y_0$, we obtain

$$y_k = \frac{y_0}{y_0 + (1 - y_0)(2 - e^{-h})^{-k}}.\qquad(4.4.5)$$

Note that

$$1 < 2 - e^{-h} < 2,\quad h > 0,\qquad(4.4.6)$$

consequently,

$$g_k = (2 - e^{-k})^{-k},\qquad(4.4.7)$$

is an exponentially decreasing function of k. Examination of Eq. (4.4.5) shows that all the solutions of Eq. (4.4.1) have the same qualitative properties as the solutions to the logistic differential equation for all step-sizes, $h > 0$.

4.4.2 ODE with Three Fixed-Points

A discrete model for Eq. (4.3.11), with a nonlocal nonlinear term, is

$$\frac{y_{k+1} - y_k}{\left(\frac{1-e^{-2k}}{2}\right)} = y_k(1 - y_{k+1}y_k).\qquad(4.4.8)$$

This expression is linear in y_{k+1}; solving for it gives

$$y_{k+1} = \frac{(1 + \bar{\phi})y_k}{1 + \bar{\phi}y_k^2},\qquad(4.4.9a)$$

where

$$\bar{\phi} = \frac{1 - e^{-2h}}{2}. \tag{4.4.9b}$$

Numerical solutions of Eq. (4.4.8) or (4.4.9) are plotted in Figure 4.4.1 for $y_0 = 0.5$ and the step-sizes $h = (0.01, 0.75, 1.5, 2.5)$. Observe that for all the selected step-sizes, the numerical solutions increase monotonically toward the limiting value of $y_\infty = \bar{y}^{(2)} = 1$. This is exactly the same qualitative behavior as the corresponding solution to the differential equation.

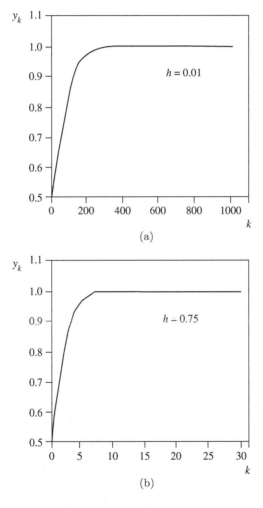

Fig. 4.4.1   Plots of Eq. (4.4.8). For each graph $y_0 = 0.5$, (a) $h = 0.01$, (b) $h = 0.75$.

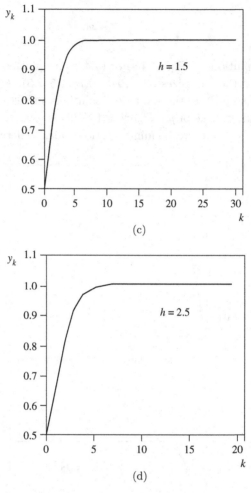

Fig. 4.4.1   Plots of Eq. (4.4.8). For each graph $y_0 = 0.5$, (c) $h = 1.5$, (d) $h = 2.5$.

For purposes of comparison, it is of interest to also examine the numerical solutions of the discrete model

$$\frac{y_{k+1} - y_k}{h} = y_k(1 - y_{k+1}y_k). \tag{4.4.10}$$

This model is constructed by using a standard forward-Euler scheme for the first-derivative and a nonlocal representation for the nonlinear term. Solving for y_k gives

$$y_{k+1} = \frac{(1+h)y_k}{1+hy_k^2}. \tag{4.4.11}$$

Figure 4.4.2 presents numerical solutions to the finite difference scheme of Eq. (4.4.10) or (4.4.11). The initial condition and step-size values are the same as in Figure 4.4.1. The obtained results can be explained by means of a linear stability analysis.

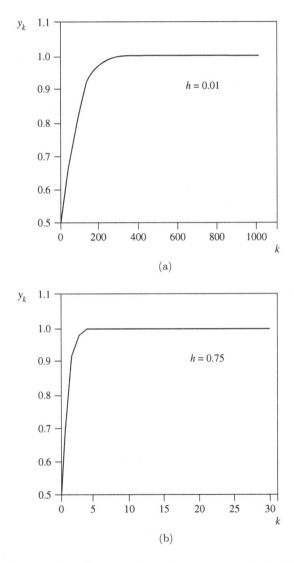

Fig. 4.4.2   Plots of Eq. (4.4.10). For each graph $y_0 = 0.5$, (a) $h = 0.01$, (b) $h = 0.75$.

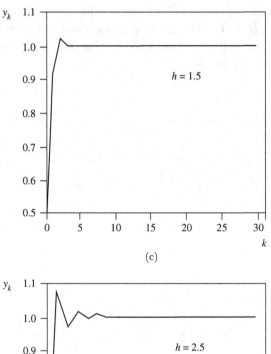

(c)

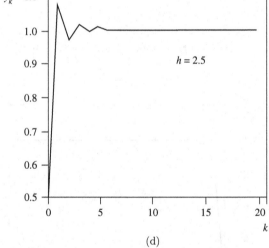

(d)

Fig. 4.4.2   Plots of Eq. (4.4.10). For each graph $y_0 = 0.5$, (c) $h = 1.5$, (d) $h = 2.5$.

Perturbations about the three fixed-points of Eq. (4.4.10) give the following linear stability equations

$$\epsilon_{k+1} = (1 + h)\epsilon_k, \tag{4.4.12}$$

$$\eta_{k+1} = \left(\frac{1 - h}{1 + h}\right)\eta_k. \tag{4.4.13}$$

(See Eq. (4.3.17).) From Eq. (4.4.12), it can be concluded that the fixed-point at $y_k = \bar{y}^{(1)} = 0$. However, the two other fixed-points, at $\bar{y}^{(2)} = 1$ and $\bar{y}^{(3)} = -1$, have the following properties:

(i) For $0 < h < 1$, the fixed-points are linearly stable.
(ii) For $h > 1$, the fixed-points are linearly stable; but, the perturbations decrease with an oscillating amplitude.

This is just the behavior seen in the various graphs of Figure 4.4.2.

4.5   Discussion

The calculations presented in this chapter show, for a scalar ordinary differential equation

$$\frac{dy}{dt} = f(y), \tag{4.5.1}$$

that the use of a renormalized denominator function

$$h \to \frac{1 - e^{-R^* h}}{R^*}, \tag{4.5.2}$$

leads to discrete models for which the fixed-points have the correct linear stability properties for all step-sizes, $h > 0$. This result is obtained whether or not a local or a nonlocal representation is used for the function $f(y)$. The procedure given for these constructions is the simplest possible for the differential equations investigated. However, more complicated discrete models exist. For example, consider the differential equation with three fixed-points

$$\frac{dy}{dt} = y(1 + y)(1 - y). \tag{4.5.3}$$

A finite difference scheme that incorporates the maximum symmetry in the nonlocal modeling of the nonlinear term is

$$\left[\frac{\Delta y_k}{\phi} - y_{k+1}(1 + y_k)(1 - y_k) \right] + \left[\frac{\Delta y_k}{\phi} - y_k(1 + y_{k+1})(1 + y_k) \right]$$
$$+ \left[\frac{\Delta y_k}{\phi} - y_k(1 + y_{k+1})(1 - y_{k+1}) \right]$$
$$- \left[\frac{\Delta y_k}{\phi} - y_k(1 + y_k)(1 - y_k) \right] = 0, \tag{4.5.4}$$

where

$$\Delta y_k = y_{k+i} - y_k, \quad \phi = \frac{1 - e^{-2k}}{2}. \tag{4.5.5}$$

(Such a form has been investigated by Price et al. [11] for an ordinary differential equation similar in form to Eq. (4.5.3). However, they consider the case where $\phi = h$.) This equation can be solved for y_{k+1} to give

$$y_{k+1} = \frac{[(5 - e^{-2h}) + (1 - e^{-2h}y_k^2]y_k}{(3 + e^{-2h}) + 3(1 - e^{-2h})y_k^2}. \tag{4.5.6}$$

A geometrical analysis [10] of Eq. (4.5.6) shows that if $y_0 > 0$, then y_k converges monotonically to the fixed-point at $\bar{y}^{(2)} = +1$. Similarly, if $y_0 < 0$, then y_k converges monotonically to the fixed-point at $\bar{y}^{(3)} = -1$. This result holds true for all $h > 0$ and corresponds exactly to the qualitative behavior of the various solutions to the differential equation. See Figures 4.5.1 and 4.5.2.

Finally, it should be emphasized that these calculations indicate that the use of a renormalized denominator function has a more important effect on the solution behavior of a discrete model than does the use of a nonlocal representation for the nonlinear term. Of course, putting both in the same discrete model produces better results.

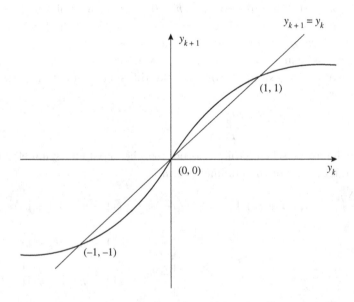

Fig. 4.5.1   Plot of Eq. (4.5.6). The fixed-points are located at $(-1, -1)$, $(0, 0)$ and $(1, 1)$.

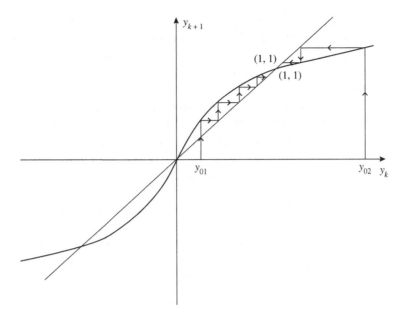

Fig. 4.5.2   Typical trajectories for Eq. (4.5.6) with $y_0 > 0$: $0 < y_{01} < 1$ and $1 < y_{02}$.

References

1. R. E. Mickens and A. Smith, *Journal of the Franklin Institute* **327**, 143–149 (1990). Finite-difference models of ordinary differential equations: Influence of denominator functions.
2. R. E. Mickens, *Dynamic Systems and Applications* 1, 329–340 (1992). Finite difference schemes having the correct linear stability properties for all finite step-sizes II.
3. R. M. Corless, C. Essex and M. A. H. Nerenberg, *Physics Letters* **A157**, 27–36 (1991). Numerical methods can suppress chaos.
4. A. Iserles, A. T. Peplow and A. M. Stuart, *SIAM Journal of Numerical Analysis* **28**, 1723–1751 (1991). A unified approach to spurious solutions introduced by time discretization.
5. R. E. Mickens, Runge-Kutta schemes and numerical instabilities: The logistic equation, in *Differential Equations and Mathematical Physics*, I. Knowles and Y. Saito, editors (Springer-Verlag, Berlin, 1987), pp. 337–341.
6. E. R. Caianiello, *Combinatorics and Renormalization in Quantum Field Theory* (W. A. Benjamin; Reading, MA; 1973).

7. G. Dahlquist, L. Edsberg, G. Skollermo and G. Soderlind, Are the numerical methods and software satisfactory for chemical kinetics?, in *Numerical Integration of Differential Equations and Large Linear Systems*, J. Hinze, editor (Springer-Verlag, Berlin, 1982), pp. 149–164.

8. L. Dieci and D. Estep, Georgia Institute of Technology, Tech. Rep. Math. 050290-039 (1990). Some stability aspects of schemes for the adaptive integration of stiff initial value problems.

9. E. N. Lorenz, *Physica* **D35**, 299–317 (1989). Computational chaos – A preclude to computational instability.

10. R. E. Mickens, *Difference Equations: Theory and Applications* (Van Nostrand Reinhold, New York, 2nd edition, 1990).

11. W. G. Price, Y. Wang and E. H. Twizell, *Numerical Methods for Partial Differential Equations* **9**, 213–224 (1993). A second-order, chaos-free, explicit method for the numerical solution of a cubic reaction problem in neurophysiology.

Chapter 5

Second-Order, Nonlinear Oscillator Equations

5.1 Introduction

Within the context of traditional classical mechanics, a one-dimensional conservative oscillator is described by a differential equation having the form [1–4]

$$\frac{d^2y}{dt^2} + f(y) = 0. \tag{5.1.1}$$

This equation has a first-integral

$$\left(\frac{1}{2}\right)\left(\frac{dy}{dt}\right)^2 + V(y) = E = \text{constant}, \tag{5.1.2}$$

where E is the constant total energy and $V(y)$, given by

$$V(y) = \int f(y)dy, \tag{5.1.3}$$

is the potential energy. Examples of several important conservative oscillator equations are the Duffing equation [3 5]

$$\frac{d^2y}{dt^2} + y + \epsilon y^3 = 0, \tag{5.1.4}$$

the modified Duffing equation

$$\frac{d^2y}{dt^2} + y^3 = 0, \tag{5.1.5}$$

and an equation arising in laser physics [6]

$$\frac{d^2y}{dt^2} + y + \frac{\epsilon y}{1 + \lambda y^2} = 0. \tag{5.1.6}$$

In these equations, ϵ and λ are generally positive constants.

A nonlinear oscillator is defined to be conservative if the differential equation is invariant under the time transformation

$$t \to -t. \tag{5.1.7}$$

The oscillator of Eq. (5.1.1) certainly has this form. However, a larger class of nonlinear conservative oscillators can be considered. For example, the equation of motion of a particle on a rotating parabola is [5]

$$\frac{d^2 y}{dt^2} + \left[\frac{1 + \epsilon(\dot{y})^2}{1 + \epsilon y^2} \right] y = 0, \tag{5.1.8}$$

where $\dot{y} = dy/dt$. This latter differential equation is also invariant under the transformation of Eq. (5.1.7). Consequently, we are led to consider a generalized conservative oscillator differential equation that has the structure

$$\frac{d^2 y}{dt^2} + f[y^2, (\dot{y})^2] y = 0. \tag{5.1.9}$$

Another important class of one-dimensional oscillators is those that have limit-cycles. In general, these systems asymptotically go to a well defined periodic state. The particular periodic state that the system finds itself in may depend on the initial conditions, but, the properties of the various periodic states are functions only of the system parameters [2–5]. The prime example of such a system is the van der Pol oscillator [2, 3, 5]

$$\frac{d^2 y}{dt^2} + y = \epsilon(1 - y^2)\frac{dy}{dt}, \quad \epsilon > 0. \tag{5.1.10}$$

This differential equation has a unique periodic solution that can be reached from any set of initial conditions in the finite $(y, \dot{y})$ phase-plane.

The purpose of this chapter is to investigate the mathematical properties of various discrete models for both conservative and limit-cycle one-dimensional oscillator differential equations. The particular equations studied will be the Duffing and van der Pol differential equations. The techniques used to construct the finite difference schemes will be based on the nonstandard modeling rules of Chapter 3. The major mathematical procedure that will be applied to obtain the analytical properties of the solutions to the difference equations is a discrete variable perturbation method formulated by Mickens [7, 8].

The books of Greenspan [9, 10] provide a good summary of his work, as well as the research of others, on the general topic of discrete physical models. Other related work on discrete-time Hamiltonian dynamics appears in references [11–14].

5.2 Mathematical Preliminaries

Consider the following class of nonlinear, second-order difference equations [8, 9]

$$\Gamma y_k = \epsilon f(y_{k+1}, y_k, y_{k-1}), \tag{5.2.1}$$

where ϵ is a parameter satisfying the condition

$$0 < \epsilon \ll 1, \tag{5.2.2}$$

and the operator Γ is defined by the relation

$$\Gamma y_k \equiv \frac{y_{k+1} - 2y_k + y_{k-1}}{4\sin^2\left(\frac{h}{2}\right)} + y_k. \tag{5.2.3}$$

We now construct a multi-discrete-variable procedure [9] to obtain perturbation solutions to Eq. (5.2.1). We begin by introducing two discrete variables k and $s = \epsilon k$ and assume that the solution to Eq. (5.2.1) has the form

$$y_k \equiv y(k, s, \epsilon) = y_0(k, s) + \epsilon y_1(k, s) + O(\epsilon^2), \tag{5.2.4}$$

where y_k is assumed to have at least a first partial derivative with respect to s. On the basis of these assumptions, we have

$$y_{k+1} = y(k+1, s+\epsilon, \epsilon) = y_0(k+1, s+\epsilon) + \epsilon y_1(k+1, s+\epsilon) + O(\epsilon^2), \tag{5.2.5}$$

$$y(k+1, s+\epsilon) = y_0(k+1, s) + \epsilon \frac{\partial y_0(k+1, s)}{\partial s} + O(\epsilon^2), \tag{5.2.6}$$

$$y_1(k+1, s+\epsilon) + y_1(k+1, s) + O(\epsilon), \tag{5.2.7}$$

and

$$y_{k+1} = y_0(k+1, s) + \epsilon \left[y_1(k+1), s) + \frac{\partial y_0(k+1, s)}{\partial s} \right] + O(\epsilon^2), \tag{5.2.8}$$

$$y_{k-1} = y_0(k-1, s) + \epsilon \left[y_1(k-1), s) - \frac{\partial y_0(k-1, s)}{\partial s} \right] + O(\epsilon^2). \tag{5.2.9}$$

Substituting Eqs. (5.2.4), (5.2.8) and (5.2.9) into Eq. (5.2.1), and setting the coefficients of the ϵ^0 and ϵ^1 terms equal to zero, gives the following determining equations for the unknown functions $y_0(k, s)$ and $y_1(k, s)$:

$$\Gamma y_0(k, s) = 0, \tag{5.2.10}$$

$$\Gamma y_1(k, s) = \frac{1}{4\sin^2\left(\frac{h}{2}\right)} \left[\frac{\partial y_0(k-1, s)}{\partial s} \right] - \left[\frac{\partial y_0(k+1, s)}{\partial s} \right]$$
$$+ f[y_0(k+1, s), y_0(k, s), y_0(k-1, s)]. \tag{5.2.11}$$

The first equation has the general solution

$$y_0(k, s) = A(s)\cos(hk) + B(s)\sin(hk),\qquad(5.2.12)$$

where, at present, $A(s)$ and $B(s)$ are unknown functions.

If Eq. (5.2.12) is substituted into the right-side of Eq. (5.2.11), then the following result is obtained ing result is obtained

$$\Gamma y_1(k, s) = \left[\lambda\frac{dA}{ds} + M_1(A, B, h)\right]\sin(hk)$$

$$+ \left[-\lambda\frac{dA}{ds} + N_1(A, B, h)\right]\cos(hk)$$

$$+ \text{(higher-order harmonics)},\qquad(5.2.13)$$

where

$$\lambda = \frac{\sin(h)}{2\sin^2\left(\frac{h}{2}\right)},\qquad(5.2.14)$$

and M_1 and N_1 are obtained from the Fourier series expansion of the function

$$f[y_0(k + 1, s), y_0(k, s), y_0(k - 1, s)] = \sum_{\ell=1}^{\infty}[M_\ell\sin(\ell hk) + N_\ell\cos(\ell hk)].$$

$$(5.2.15)$$

The "higher-order harmonics" term in Eq. (5.2.13) is the sum on the right-side of Eq. (5.2.15) for $\ell > 2$. If $y_1(ks)$ is to be bounded, i.e., contain no secular terms, then the coefficients of the $\sin(hk)$ and $\cos(hk)$ terms must be zero [8]. This condition gives equations that can be solved to get the functions $A(s)$ and $B(s)$; they are

$$\lambda\frac{dA}{ds} + M_1(A, B, h) = 0,\qquad(5.2.16)$$

$$\lambda\frac{dB}{ds} - N_1(A, B, h) = 0.\qquad(5.2.17)$$

Substitution of $A(s)$ and $B(s)$ into Eq. (5.2.12) provides a uniformly valid solution to Eq. (5.2.1), up to terms of order ϵ.

5.3   Conservative Oscillators

The periodic solutions of conservative oscillators have the property that the amplitude of the oscillations are constant [1, 2, 3, 4]. In this section, we use this property as the characteristic defining a conservative oscillator. Our interest is in applying this criterion to the solutions of various discrete models of conservative oscillators. Without loss of generality, we only consider the Duffing equation [15, 16]

$$\frac{d^2y}{dt^2} + y + \epsilon y^3 = 0, \quad 0 < \epsilon \ll 1. \tag{5.3.1}$$

For $\epsilon > 0$, it can be shown that all the solutions to the Duffing differential equation are bounded and periodic [1, 2]. For small ϵ, a uniformly valid expression for $y(t)$ can be calculated by using a variety of perturbation procedures [3, 5]. They all give, to terms of order ϵ, the result

$$y(t) = \bar{a} \sin\left[\left(1 + \frac{3\bar{a}^2\epsilon}{8}\right)t + \phi\right], \tag{5.3.2}$$

where $\bar{a}$ is the constant amplitude and ϕ the constant phase.

We now construct and analyze four discrete models of the Duffing equation for the parameter domain where $0 < \epsilon \ll 1$. Thus, the discrete-multi-time perturbation method of the previous section can be applied. Any finite difference scheme that has solutions for which the amplitude is not constant will be considered to be an inappropriate discrete model [16].

The four finite difference models of Eq. (5.3.1) to be investigated are

$$\Gamma y_k + \epsilon y_k^3 = 0, \tag{5.3.3}$$

$$\Gamma y_k + \epsilon y_k^2 \left(\frac{y_{k+1} + y_{k-1}}{2}\right) = 0, \tag{5.3.4}$$

$$\Gamma y_k + \epsilon y_{k-1}^3 = 0, \tag{5.3.5}$$

$$\Gamma y_k + \epsilon y_{k+1}^3 = 0, \tag{5.3.6}$$

where the operator Γ is defined by Eq. (5.2.3). Note that for $\epsilon = 0$, the linear finite difference scheme obtained

$$\Gamma y_k = 0, \tag{5.3.7}$$

is an exact discrete model for the harmonic oscillator differential equation

$$\frac{d^2y}{dt^2} + y = 0. \tag{5.3.8}$$

We now present, in detail, the calculations for the perturbation solution to Eq. (5.3.3). The other three examples are done in exactly the same manner.

Comparison of Eqs. (5.2.1) and (5.3.3) gives

$$f(y_{k+1}y_k, y_{k-1}) = -y_k^3. \tag{5.3.9}$$

Using the result

$$[y_0(k, s)]^3 = [A\cos(h) + B\sin(hk)]^3 = \left(\frac{3}{4}\right) A(A^2 + B^2)\cos(hk)$$

$$+ \left(\frac{3}{4}\right) B(A^2 + B^2)\sin(hk) + + \left(\frac{3}{4}\right) B(A^2 - B^2)\cos(3hk)$$

$$+ \left(\frac{3}{4}\right) B(3A^2 - B^2)\sin(3hk), \tag{5.3.10}$$

we find

$$M_1 = -\left(\frac{3}{4}\right) B(A^2 + B^2), \tag{5.3.11}$$

$$N_1 = -\left(\frac{3}{4}\right) A(A^2 + B^2), \tag{5.3.12}$$

and

$$\lambda\frac{dA}{ds} = \left(\frac{3}{4}\right) B(A^2 + B^2), \tag{5.3.13}$$

$$\lambda\frac{dB}{ds} = -\left(\frac{3}{4}\right) A(A^2 + B^2), \tag{5.3.14}$$

where λ is defined in Eq. (5.2.14). Multiplying Eqs. (5.3.13) and (5.3.14), respectively, by A and B, and adding the resulting expressions gives

$$\frac{d}{ds}(A^2 + B^2) = 0, \tag{5.3.15}$$

or

$$\bar{a}^2 = A^2 + B^2 = \text{constant.} \tag{5.3.16}$$

Now define ω to be

$$\omega = \frac{3\bar{a}^2}{4\lambda}. \tag{5.3.17}$$

Then Eqs. (5.3.13) and (5.3.14) become

$$\frac{dA}{ds} = \omega B, \quad \frac{dB}{ds} = -\omega A. \tag{5.3.18}$$

They have solutions

$$A(s) = \bar{a}\sin(\omega s + \phi), \tag{5.3.19a}$$

$$B(s) = \bar{a}\cos(\omega s + \phi), \tag{5.3.19b}$$

where ϕ is an arbitrary constant. If these results are substituted into Eq. (5.2.12), then the following is obtained

$$y_0(k, s) = \bar{a} \sin(\omega s + hk + \phi). \tag{5.3.20}$$

Now using $s = \epsilon k$, we finally obtain

$$y_k = y_0(k, s) + O(\epsilon) = \bar{a} \sin\left\{1 + \left(\frac{3\bar{a}^2\epsilon}{2}\right)\left[\frac{\sin^2(h/2)}{h\sin(h)}\right]e_k + \phi\right\}, \tag{5.3.21}$$

where $t_k = hk$. Note that in the limits

$$h \to 0, \quad k \to \infty, \quad hk = t = \text{fixed}, \tag{5.3.22}$$

the right-side of Eq. (5.3.21) tends to the function of Eq. (5.3.2) as expected.

The significant point is that to first-order in ϵ, the discrete model of the Duffing equation, given by Eq. (5.3.3), has oscillatory solutions with constant amplitude. Thus, using only this criterion, the finite difference scheme of Eq. (5.3.3) is an adequate model. Note that Eq. (5.3.3) uses a standard local expression for the nonlinear term y^3, i.e.,

$$y^3 \to y_k^3. \tag{5.3.23}$$

However, the linear part of the differential equation is modeled by its exact finite difference scheme.

Now let $\bar{a}_i$, where $i = (1, 2, 3, 4)$, be the amplitudes of the solutions, respectively, of Eqs. (5.3.3) to (5.3.6). We have just obtained $\bar{a}_1$. Repeating the calculation for the other three cases gives

$$\frac{d}{ds}(\bar{a}_1)^2 = 0, \tag{5.3.24}$$

$$\frac{d}{ds}(\bar{a}_2)^2 = 0, \tag{5.3.25}$$

$$\frac{d}{ds}(\bar{a}_3)^2 = \alpha(\bar{a}_3)^4, \tag{5.3.26}$$

$$\frac{d}{ds}(\bar{a}_4)^2 = -\alpha(\bar{a}_4)^4, \tag{5.3.27}$$

where

$$\alpha = 3\sin^2\left(\frac{h}{2}\right). \tag{5.3.28}$$

Observe that the finite difference schemes of Eqs. (5.3.3) and (5.3.4) give oscillatory solutions for which the amplitude is constant. However, the discrete models of Eqs. (5.3.5) and (5.3.6) have oscillatory solutions whose amplitudes, respectively, increase and decrease, a behavior not consistent with the known properties of the Duffing differential equation. Therefore, we conclude that these discrete models are not appropriate for calculating numerical solutions.

Further examination of Eqs. (5.3.3) and (5.3.4) shows that they have the special symmetry

$$y_{k+1} \leftrightarrow y_{k-1}. \tag{5.3.29}$$

This result can be generalized in the following manner. Let

$$\Gamma y_k + F(y_{k+1}, y_k, y_{k-1}) = 0, \tag{5.3.30}$$

be a discrete model of the conservative oscillator differential equation

$$\frac{d^2 y}{dt^2} + f(y) = 0; \tag{5.3.31}$$

then Eq. (5.3.30) will be an appropriate discrete model if the function $F(y_{k+1}, y_k, y_{k+1})$ has the property

$$F(y_{k+1}, y_k, y_{k+1}) = F(y_{k-1}, y_k, y_{k+1}). \tag{5.3.32}$$

The existence of this symmetry can be easily understood in terms of the time-reversal invariance of the conservative differential equation. The linear part of the discrete model, Γy_k, automatically satisfies this condition.

There is also a second way of investigating the conservative nature of finite difference models of differential equations, namely, the use of Taylor series expansions in h to determine the governing differential equation for fixed, but, nonzero values of the step-size. We now demonstrate how this can be accomplished.

In the limits, given by Eq. (5.3.22), all of the above discrete models of the Duffing equation converge [17] to the differential equation. However, in practical calculations, i.e., the numerical integration of the Duffing differential equation, h may be small, but is nevertheless always finite. For this situation, the discrete model corresponds to an entirely different differential equation. To demonstrate this, we make use of the following results:

$$y_k = y(t) + O(h^2), \tag{5.3.33a}$$

$$y_{k\pm 1} = y(t) \pm h\frac{dy}{dt} + O(h^2), \tag{5.3.33b}$$

$$\frac{y_{k+1} - 2y_k + y_{k-1}}{4\sin^2\left(\frac{h}{2}\right)} = \frac{d^2 y}{dt^2} + O(h^2), \tag{5.3.33c}$$

$$(y_{k\pm1})^3 = y^3 \pm 3hy^2 \left(\frac{dy}{dt}\right) + O(h^2). \tag{5.3.33d}$$

Substituting Eqs. (5.3.33) into Eqs. (5.3.3) to (5.3.6) gives, respectively,

$$\frac{d^2y}{dt^2} + y + \epsilon y^3 = O(h^2), \tag{5.3.34}$$

$$\frac{d^2y}{dt^2} + y + \epsilon y^3 = O(h^2), \tag{5.3.35}$$

$$\frac{d^2y}{dt^2} - 3\epsilon hy^2 \left(\frac{dy}{dt}\right) + y + \epsilon y^3 = O(h^2), \tag{5.3.36}$$

$$\frac{d^2y}{dt^2} + 3\epsilon hy^2 \left(\frac{dy}{dt}\right) + y + \epsilon y^3 = O(h^2). \tag{5.3.37}$$

Examination of these equations shows that to terms of order h^2, the discrete models given by Eqs. (5.3.3) and (5.3.4) represent the conservative Duffing oscillator differential equation. However, the discrete models of Eqs. (5.3.5) and (5.3.6) correspond, to terms of order h^2, respectively, oscillators with negative and positive damping. This is exactly the result found above using the discrete-two-time perturbation method. Similar results hold for the linear harmonic oscillator differential equation; see Section 2.3. The symmetric replacement $y^3 \to y_{k+1}y_ky_{k-1}$ also gives Eqs. (5.3.34) and (5.3.35).

Consider again the generalized conservative oscillator differential equation

$$\frac{d^2y}{dt^2} + g[y^2, (\dot{y})^2] = 0, \tag{5.3.38}$$

where $\dot{y} \equiv dy/dt$. In practical applications, the function $g[y^2, (\dot{y})^2]$ is either a polynomial function or a rational function whose denominator has no real zeros. A conservative discrete model for Eq. (5.3.38) can be obtained by replacing the second-derivative with

$$\frac{d^2y}{dt^2} \to \frac{y_{k+1} - 2y_k + y_{k-1}}{\psi(h)}, \tag{5.3.39a}$$

where

$$\psi(h) = h^2 + O(h^4), \tag{5.3.39b}$$

and replacing the nonlinear function g by a function

$$g[y^2, (\bar{y})^2]y \to G(y_{k+1}, y_k, y_{k-1}) \tag{5.3.40}$$

having the following two properties

$$G(y_{k+1}, y_k, y_{k-1}) = G(y_{k+1}, y_k, y_{k+1}), \tag{5.3.41a}$$

$$G(y_{k+1}, y_k, y_{k-1}) = g[y^2, (\dot{y})^2]y + O(h^2). \tag{5.3.41b}$$

A particular way to implement this is to replace in $g[y^2, (\dot{y})^2]y$, the y^2 and $(\dot{y})^2$ terms, respectively, by some linear combinations of expressions such as

$$y^2 \rightarrow \begin{cases} y_k^2 \\ \frac{y_{k+1}^2 + y_k^2 + y_{k-1}^2}{3}, \\ \frac{y_k(y_{k+1} + y_{k-1})}{2}, \\ y_{k+1}y_{k-1}, \\ \text{etc.,} \end{cases} \tag{5.3.42}$$

$$(\dot{y})^2 \rightarrow \begin{cases} \left[\frac{y_{k+1} - y_{k-1}}{2\psi(h)}\right]^2, \\ \left(\frac{1}{2}\right)\left[\frac{y_{k+1} - y_k}{\psi(h)}\right]^2 + \left(\frac{1}{2}\right)\left[\frac{y_k - y_{k-1}}{\psi(h)}\right]^2, \\ \text{etc.,} \end{cases} \tag{5.3.43}$$

where

$$\psi(h) = h + O(h^2). \tag{5.3.44}$$

Classical mechanics defines a conservative oscillator as one for which there exists a constant energy first-integral and all bounded solutions are periodic [1, 2]. Earlier, in Section 3.5, we discussed a best finite difference scheme for the Duffing oscillator. There our interest was in constructing a discrete model that had a constant (discrete) first integral. We showed that this was possible for the scheme

$$\frac{y_{k+1} - 2y_k + y_{k-1}}{\psi} + y_k + \epsilon(y_k)^2 \left(\frac{y_{k+1} + y_{k-1}}{2}\right) = 0, \tag{5.3.45}$$

where

$$\psi = h^2 + O(h^4), \tag{5.3.46}$$

but, not possible for

$$\frac{y_{k+1} - 2y_k + y_{k-1}}{\psi} + y_k + \epsilon(y_k)^2 = 0. \tag{5.3.47}$$

In this section, an alternative definition of a conservative oscillator was introduced, namely, a second-order differential equation that is invariant under the transformation $t \leftrightarrow -t$. This invariance translates to the requirement that the discrete model should be unchanged when $y_{k+1} \leftrightarrow y_{k-1}$ with the consequence that the amplitude of oscillatory solutions be constant. With this alternative definition, both of the discrete models given by Eqs. (5.3.45) and (5.3.47) are conservative. From the view point of the actual physical phenomena, only the scheme of Eq. (5.3.45) can be used, since the oscillator does in fact satisfy an energy conservation law. This scheme is the one that comes from the application of the nonstandard modeling rules given in Section 3.4.

Finally, for comparison with the first-order in ϵ solution of Eq. (5.3.3), given by Eq. (5.3.21), we give the solution of Eq. (5.3.4). It is

$$y_k = y_0(k, s) + O(\epsilon) = \bar{a}\sin[(1 + \epsilon\alpha)t_k + \phi] + O(\epsilon), \qquad (5.3.48)$$

where

$$\alpha = \left(\frac{3\bar{a}^2}{2}\right) 2 \left[\frac{\cos(h)\sin^2(h/2)}{h\sin(h)}\right], \qquad (5.3.49)$$

and $t_k = hk$.

5.4 Limit-Cycle Oscillators

The nonlinear differential equation of van der Pol [2, 3, 5] serves as an important model equation for one-dimensional dynamic systems having a single, stable limit-cycle. After constructing four finite difference models, we will use a discrete multitime perturbation procedure to calculate solutions to the finite difference equations. A detailed comparison will then be made between these solutions and the corresponding solution of the van der Pol differential equation [18]. One of the issues to be considered is what discrete form should be used to model the first-derivative [19]?

The van der Pol oscillator differential equation is

$$\frac{d^2y}{dt^2} + y = \epsilon(1 - y^2)\frac{dy}{dt}, \quad \epsilon > 0. \qquad (5.4.1)$$

For the purposes of the present section, we assume that ϵ satisfies the condition

$$0 < \epsilon \ll 1. \qquad (5.4.2)$$

Under this restriction, the limit-cycle is given by the expression [3]

$$y(\theta) = 2\cos\theta + \left(\frac{\epsilon}{4}\right)(3\sin\theta - \sin 3\theta)$$

$$+ \left(\frac{\epsilon^2}{96}\right)(-13\cos\theta + 18\cos 3\theta - 5\cos 5\theta) + O(\epsilon^3), \qquad (5.4.3)$$

where

$$\theta = \left[1 - \frac{\epsilon^2}{16} + O(\epsilon^3)\right] t. \tag{5.4.4}$$

Near the limit-cycle, the solution is [3]

$$y(t) = \frac{2a_0 \cos(t + \phi_0)}{[a_0^2 - (a_0^2 - 4)\exp(-\epsilon t)]^{1/2}} + O(\epsilon), \tag{5.4.5}$$

where a_0 and ϕ are constants. Note that

$$\lim_{t \to \infty} y(t) = 2\cos t + O(\epsilon). \tag{5.4.6}$$

The four discrete models to be investigated are
Model I-A:

$$\Gamma y_k = \epsilon(1 - y_k^2)\left[\frac{y_{k+1} - y_{k-1}}{2h}\right]; \tag{5.4.7}$$

Model I-B:

$$\Gamma y_k = \epsilon(1 - y_k^2)\left[\frac{y_{k+1} - y_{k-1}}{2\left(\frac{4\sin^2(h/2)}{h}\right)}\right]; \tag{5.4.8}$$

Model II:

$$\Gamma y_k = \epsilon\left\{1 - y_k\left[\frac{y_{k+1} + y_{k-1}}{2\cos(h)}\right]\right\}\left[\frac{y_{k+1} - y_{k-1}}{2\sin(h)}\right]; \tag{5.4.9}$$

Model III:

$$\Gamma y_k = \epsilon(1 - y_k^2)\left[\frac{y_k - y_{k-1}}{\left(\frac{4\sin^2(h/2)}{h}\right)}\right]; \tag{5.4.10}$$

where the operator Γ is defined by Eq. (5.2.3). Note that Models I-A, I-B and II use a central difference for the first-derivative, while Model III expresses the first-derivative, by a backward-Euler. Also, the denominator functions for all but Model I-A are nonstandard. Model II has been studied previously by Potts [20].

Application of the discrete multi-time perturbation method to Eqs. (5.4.7), (5.4.8), (5.4.9) and (5.4.10) gives the following results:
Model I-A:

$$y_k = \frac{2a_0 \cos(t_k + \phi_0)}{[a_0^2 - (a_0^2 - 4)\exp(-\lambda_1 \epsilon t_k)]^{1/2}} + O(\epsilon), \tag{5.4.11a}$$

$$\lambda_1 = \frac{4\sin^2(h/2)}{h^2}; \tag{5.4.11b}$$

Model I-B:

$$y_k = \frac{2a_0 \cos(t_k + \phi_0)}{[a_0^2 - (a_0^2 - 4) \exp(-\epsilon t_k)]^{1/2}} + O(\epsilon); \qquad (5.4.12)$$

Model II:

$$y_k = \frac{2a_0 \cos(t_k + \phi_0)}{[a_0^2 - (a_0^2 - 4) \exp(-\lambda_2 \epsilon t_k)]^{1/2}} + O(\epsilon); \qquad (5.4.13a)$$

$$\lambda_2 = \frac{4 \sin^2(h/2)}{h^2}; \qquad (5.4.13b)$$

Model III:

$$y_k = \frac{2a_0 \cos(t_k + \phi_0)}{[a_0^2 - (a_0^2 - 4) \exp(-\epsilon t_k)]^{1/2}} + O(\epsilon). \qquad (5.4.14)$$

To illustrate how the above results were obtained, we show the details of the calculation for Model I-A. From Section 5.2, the function $f(y_{k+1}, y_k, y_{k-1})$ is

$$f = (1 - y_k^2) \left[\frac{y_{k+1} - y_{k-1}}{2h} \right]. \qquad (5.4.15)$$

From this, the amplitude functions $A(s)$ and $B(s)$ are determined by the equations

$$\frac{dA}{ds} = - \left[\frac{\sin^2(h/2)}{2h} \right] A(A^2 + B^2 - 4), \qquad (5.4.16)$$

$$\frac{dB}{ds} = - \left[\frac{\sin^2(h/2)}{2h} \right] B(A^2 + B^2 - 4). \qquad (5.4.17)$$

Define z as

$$z = A^2 + B^2,$$

multiply Eqs. (5.4.16) and (5.4.17), respectively, by A and B, and add the resulting expressions to obtain the following result

$$\frac{dz}{ds} = - \frac{\sin^2(h/2)}{h} z(z - 4). \qquad (5.4.18)$$

Let $z(0) = z_0$, then Eq. (5.4.18) has the solution

$$z(s) = \frac{4z_0}{z_0 - (z_0 - 4) \exp(-\bar{\lambda}_1 s)}, \qquad (5.4.19)$$

$$\bar{\lambda}_1 = \frac{4 \sin^2(h/2)}{h}. \qquad (5.4.20)$$

Since $s = \epsilon k$ and $t_k = hk$, then

$$\bar{\lambda}_1 s = \lambda_1 \epsilon t_k, \tag{5.4.21}$$

where

$$\lambda_1 = \frac{4\sin^2(h/2)}{h^2} \tag{5.4.22}$$

Now Eq. (5.2.12) can be rewritten to the form

$$y_0(k, s) = A(s)\cos(t_k) + B(s)\sin(t_k) = a(s)\cos[t_k + \phi(s)], \tag{5.4.23}$$

where

$$a^2 = A^2 + B^2, \quad \tan\phi = -\frac{B}{A}. \tag{5.4.24}$$

From Eqs. (5.4.16) and (5.4.17), it follows that

$$\frac{dA}{dB} = \frac{A}{B}, \tag{5.4.25}$$

or

$$A(s) = cB(s), \tag{5.4.26}$$

where c is an arbitrary constant. This shows that the phase $\phi(s)$ is a constant, i.e.,

$$\phi(s) = \phi_0 = \text{constant}. \tag{5.4.27}$$

Putting all these results together gives Eqs. (5.4.11).

Examination of the four discrete models for the van der Pol equation indicates that under the condition of Eq. (5.4.2), Models I-B and III give the same result as the perturbation solution to the van der Pol differential equation. However, for all four models, we have the proper convergence, i.e.,

$$\underset{\substack{h \to 0 \\ k \to \infty \\ hk=t=\text{fixed}}}{\text{Lim}} \quad y_k = y(t). \tag{5.4.28}$$

Note that both Models I-B and III use nonstandard denominator functions for the discrete first-derivative.

The question now arises: Can a principle or requirement be formulated that will allow us to select either Model I-B or Model III? The answer to this question is yes. The requirement will be that the lowest order term in the Taylor series expansion (in h) of the discrete model should reproduce the original van der Pol differential equation. We used this technique in

Section 5.3 to explain the behavior of various discrete models of the Duffing equation.

Let $\phi(h)$ be defined as

$$\psi(h) \equiv \frac{4\sin^2(h/2)}{h}. \tag{5.4.29}$$

An elementary calculation gives

$$\Gamma y_k = \frac{d^2y}{dt^2} + y + O(h^2), \tag{5.4.30}$$

$$1 - y_k^2 = 1 - y^2 + O(h^2), \tag{5.4.31}$$

$$\frac{y_{k+1} - y_{k-1}}{2\psi(h)} = \frac{dy}{dt} + O(h^2), \tag{5.4.32}$$

$$\frac{y_k - y_{k-1}}{\psi(h)} = \frac{dy}{dt} - \left(\frac{h}{2}\right)\frac{d^2y}{dt^2} + O(h^2). \tag{5.4.33}$$

From these expressions, it follows that:

Model I-B:

$$\Gamma y_k - \epsilon(1 - y_k^2)\left[\frac{y_{k+1} - y_{k-1}}{2\psi(h)}\right] = \left[\frac{d^2y}{dt^2} + y - \epsilon(1 - y^2)\frac{dy}{dt}\right] + O(h^2); \tag{5.4.34}$$

Model III:

$$\Gamma y_k - \epsilon(1 - y_k^2)\left[\frac{y_k - y_{k-1}}{\psi(h)}\right]$$
$$= \left[\frac{d^2y}{dt^2} + y - \epsilon(1 - y^2)\frac{dy}{dt} + \left(\frac{\epsilon h}{2}\right)(1 - y^2)\frac{d^2y}{dt^2}\right] + O(h^2). \tag{5.4.35}$$

Note that to $O(h^2)$, Model I-B gives the van der Pol differential equation. However, this does not occur for Model III since there is an $O(h)$ term. Thus, the finite difference scheme of Model III describes to $O(h^2)$ the modified van der Pol equation

$$\frac{d^2y}{dt^2} + y = \epsilon(1 - y^2)\frac{dy}{dt} - \left(\frac{\epsilon h}{2}\right)(1 - y^2)\frac{d^2y}{dt^2}. \tag{5.4.36}$$

Using the method of harmonic balance [21], the influence of the extra term on the right-side of Eq. (5.4.36) can be determined. The approximate solution to the limit-cycle of the van der Pol equation is [21]

$$y_{\text{VDP}}(t) = 2\cos t. \tag{5.4.37}$$

The same calculation applied to the above modified van der Pol equation gives

$$y_{\text{MVDP}}(t) = 2\cos\left[\left(1 + \frac{\epsilon h}{2}\right)t\right]. \tag{5.4.38}$$

Note that besides a small shift in the frequency, the modified van der Pol equation has essentially the same properties as the usual van der Pol oscillator. However, because of the extra term that Model III introduces, our preference would be to select Model I-B as the finite difference scheme to use in the numerical integration of the van der Pol differential equation.

5.5   General Oscillator Equations

A large class of one-dimensional, nonlinear oscillators can be modeled by differential equations having the form [1–6]

$$\frac{d^2y}{dt^2} + y + f(y^2)\frac{dy}{dt} + g(y^2)y = 0. \tag{5.5.1}$$

For example, the Duffing equation corresponds to

$$f(y^2) = 0, \quad g(y^2) = \epsilon y^2, \tag{5.5.2}$$

while the van der Pol equation has

$$f(y^2) = -\epsilon(1 - y^2), \quad g(y^2) = 0. \tag{5.5.3}$$

The results of the previous two sections suggest the following nonstandard finite difference scheme for Eq. (5.5.1)

$$\frac{y_{k+1} - 2y_k + y_{k-1}}{4\sin^2\left(\frac{h}{2}\right)} + y_k + f(y_k^2)\left[\frac{\frac{y_{k+1} - y_{k-1}}{4\sin^2\left(\frac{h}{2}\right)}}{h}\right]$$
$$+ g(y_k^2)\left(\frac{y_{k+1} + y_{k-1}}{2}\right) = 0. \tag{5.5.4}$$

Note that Eq. (5.5.4) has the features:

(i) When $f(y^2) = 0$ and $g(y^2) = 0$, the discrete model is an exact finite difference scheme for the harmonic oscillator equation

$$\frac{d^2y}{dt^2} + y = 0. \tag{5.5.5}$$

(ii) If $f(y^2) = 0$, then Eq. (5.5.1) is the equation of motion for a conservative oscillator. The discrete form for $g(y^2)y$ is constructed on the results obtained in Section 5.3 for conservative oscillators.

(iii) The form for $g(y^2)y = 0$ is consistent with the analysis of the van der Pol oscillator as discussed in Section 5.4.

(iv) Of importance is that the discrete model given by Eq. (5.5.4) is explicit, i.e., y_{k+2} can be solved for and expressed in terms of y_k and y_{k+1}. Consequently, a numerical evaluation of Eq. (5.5.4) involves only a two-step iteration procedure, i.e., shifting the index by one and solving Eq. (5.5.4) for

$$y_{k+2} = \frac{[2\cos(h)]y_{k+1} - [1 - hf(y_{k+1}^2) + 4\sin^2\left(\frac{h}{2}\right)g(y_{k+1}^2)]y_k}{1 + hf(y_{k+1}^2) + 4\sin^2\left(\frac{h}{2}\right)g(y_{k+1}^2)}. \qquad (5.5.6)$$

5.6 Response of a Linear System [22]

The linear, damped oscillator can be used to model a large number of systems in the physical [1, 4] and engineering sciences [23, 24]. In particular, such a class of systems arises in civil earthquake engineering [24–26]. For a one-dimensional system, the equation of motion takes the form

$$\frac{d^2y}{dt^2} + 2c\omega\frac{dy}{dt} + \omega^2 y = g(t), \qquad (5.6.1)$$

where ω is the angular frequency of the undamped system, c is related to the damping coefficient, and $g(t)$ is a forcing function. In many applications, the forcing function is measured or known only at fixed, equal time intervals, i.e.,

$$g(t) \to g_k, \quad t_k = hk, \qquad (5.6.2)$$

where k is an integer and $h = \Delta t$ is the interval between measurement of $g(t)$. Thus, with this limitation in mind, a discrete form of the left-side is required to have Eq. (5.6.1) make physical sense. This section gives a general computational procedure for calculating "$y(t)$" [22] and generalizes the work presented by Ly [27].

To proceed, we rewrite Eq. (5.6.1) in the dimensionless form [3]

$$\frac{d^2y}{dt^2} + 2\epsilon\frac{dy}{dt} + y = f(t), \qquad (5.6.3)$$

where ϵ is the dimensionless damping constant and t is a dimensionless time variable. Selection of the initial conditions

$$y(t_0) = y_0, \quad \dot{y}(t_0) = \dot{y}_0, \qquad (5.6.4)$$

where the "dots" indicate time derivatives, we can express the general solution to Eqs. (5.6.3) and (5.6.4) as [28]

$$y(t) = M(t)[(\beta \cos \beta t_0 - \epsilon \sin \beta t_0)y_0 - (\sin \beta t_0)\dot{y}_0] \cos \beta t$$
$$+ M(t)[(\epsilon \cos \beta t_0 + \beta \sin \beta t_0)y_0 + (\cos \beta t_0)\dot{y}_0] \sin \beta t$$
$$+ N(t) \int_{t_0}^{t} f(s)e^{ts}[(\sin \beta t) \cos \beta s - (\cos \beta t) \sin \beta s]ds, \qquad (5.6.5)$$

where

$$\begin{cases} M(t) = \exp[-\epsilon(t - t_0)]/\beta, \\ N(t) = \exp(-\epsilon t)/\beta, \\ \beta^2 = 1 - \epsilon^2. \end{cases} \qquad (5.6.6)$$

The derivative, $\dot{y}(t)$, can be found by differentiating Eq. (5.6.5) using the results of Eq. (5.6.6).

If $y_1(t)$ and $y_2(t)$ are defined to be

$$y_1(t) \equiv y(t), \quad y_2(t) \equiv \dot{y}(t), \qquad (5.6.7)$$

then Eq. (5.6.3) takes the form

$$\frac{dy_1}{dt} = y_2, \qquad (5.6.8)$$

$$\frac{dy_2}{dt} = -2\epsilon y_2 - y_1 + f(t). \qquad (5.6.9)$$

The exact finite difference scheme [29] for this system of first-order differential equations finite difference can be determined in a straightforward, but, long calculation. It is [22]

$$y_1(k + 1) = R[\beta \cos \beta h + \epsilon \sin \beta h]y_1(k) + R[\sin \beta h]y_2(k)$$
$$+ S[\sin \beta h(k + 1)] \int_{hk}^{h(k+1)} f(s)e^{\epsilon s} \cos \beta s ds$$
$$- S[\cos \beta h(k + 1)] \int_{hk}^{h(k+1)} f(s)e^{\epsilon s} \sin \beta s ds, \qquad (5.6.10)$$

$$y_2(k + 1) = R[\sin \beta h]y_1(k) + R[-\epsilon \sin \beta h + \beta \cos \beta h]y_2(k)$$
$$+ S[\beta \cos \beta h(k + 1) - \epsilon \sin \beta h(k + 1)] \int_{hk}^{h(k+1)} f(s)e^{\epsilon s} \sin \beta s ds$$
$$+ S[\epsilon \cos \beta h(k + 1) + \beta \sin \beta h(k + 1)] \int_{hk}^{h(k+1)} f(s)e^{\epsilon s} \sin \beta s ds,$$
$$(5.6.11)$$

where

$$\begin{cases} R = \exp(-\epsilon h)/\beta, \\ S = \exp[-\epsilon h(k+1)]/\beta, \end{cases} \tag{5.6.12}$$

and

$$y_1(k) = y_1(t_k), \quad y_2(k) = y_2(t_k). \tag{5.6.13}$$

Let the forcing function, $f(t)$, be known only at the discrete times $t_k = hk$. What should we do in this situation? The simplest way to proceed is to replace $f(s)$ in the integrands of Eqs. (5.6.10) and (5.6.11) by a linear functional form in each time interval t_k to t_{k+1}. This gives essentially the results obtained by Ly [27]. However, if additional information is available or if a nonlinear variation of f is used in each time interval [30], then Eqs. (5.6.10) and (5.6.11) determine the corresponding discrete model once the integrals are evaluated. Also, the acceleration, can be calculated from

$$\ddot{y}(t) \equiv y_3(t) = -2\epsilon \dot{y}(t) - y(t) + f(t). \tag{5.6.14}$$

Therefore, $\ddot{y}(t)$ can be determined from a knowledge of $y_1(k)$ and $y_2(k)$, by using the relation

$$y_3(k) = -2\epsilon y_2(k) - y_1(k) + f_k. \tag{5.6.15}$$

Finally, the finite difference scheme constructed above should be stable since it was obtained from the exact discrete model of the unforced oscillator.

References

1. H. Goldstein, *Classical Mechanics* (Addison-Wesley; Reading, MA; 1980).
2. J. J. Stoker, *Nonlinear Vibrations* (Interscience, New York, 1987).
3. R. E. Mickens, *Nonlinear Oscillations* (Cambridge University Press, New York, 1981).
4. A. B. Pippard, *The Physics of Vibration I* (Cambridge University Press, Cambridge, 1978).
5. A. H. Nayfeh, *Problems in Perturbation* (Wiley-Interscience, New York, 1985).
6. N. Bessis and G. Bessis, *Journal of Mathematical Physics* **21**, 2780–2791 (1980). A note on the Schrodinger equation for the $x^2 + \lambda x^2/(1 + gx^2)$ potential.

7. R. E. Mickens, *Journal of the Franklin Institute* **321**, 39–47 (1986). Periodic solutions of second-order nonlinear difference equations containing a small parameter III. Perturbation theory.

8. R. E. Mickens, *Journal of Franklin Institute* **324**, 263–271 (1987). Periodic solutions of second-order nonlinear difference equations containing a small parameter IV. Multi-discrete time method.

9. D. Greenspan, *Discrete Models* (Addison-Wesley; Reading, MA; 1973).

10. D. Greenspan, *Arithmetic Applied Mathematics* (Pergamon, New York, 1980).

11. Y. Shibbern, *Discrete-Time Hamiltonian Dynamics* (Ph. D. Thesis, University of Texas at Arlington, 1992).

12. R. Labudde, *International Journal of General Systems* **6**, 3–12 (1980). Discrete Hamiltonian mechanics.

13. T. D. Lee, *Journal of Statistical Physics* **46**, 843–860 (1987). Difference equations and conservation laws.

14. Y. Wu, *Computers and Mathematics with Applications* **20**, 61–75 (1990). The discrete variational approach to the Euler-Lagrange equations.

15. R. E. Mickens, O. Oyedeji and C. R. McIntyre, *Journal of Sound and Vibration* **130**, 509–512 (1989). A difference equation model of the Duffing equation.

16. R. E. Mickens, *Journal of Sound and Vibration* **124**, 194–198 (1988). Properties of finite difference models of non-linear conservative oscillators.

17. H. J. Stetter, *Analysis of Discretization Methods for Ordinary Differential Equations* (Springer-Verlag, Berlin, 1973).

18. R. E. Mickens, Investigation of finite-difference models of the van der Pol equation, in *Differential Equations and Applications*, A. R. Aftabizadeh, editor (Ohio University Press; Columbus, OH; 1988), pp. 210–215.

19. R. E. Mickens and A. Smith, *Journal of the Franklin Institute* **327**, 143–149 (1990). Finite-difference models of ordinary differential equations: Influence of denominator functions.

20. R. B. Potts, *Nonlinear Analysis* **7**, 801–812 (1983). Van der Pol difference equation.

21. R. E. Mickens, *Journal of Sound and Vibration* **94**, 456–460 (1984). Comments on the method of harmonic balance.

22. R. E. Mickens, *Journal of Sound and Vibration* **112**, 183–186

(1987). A computational method for the determination of the response of a linear system.

23. L. A. Pipes and L. R. Harvill, *Applied Mathematics for Engineers and Physicists* (McGraw-Hill, New York, 1970).

24. R. W. Clough and J. Penzien, *Dynamics of Structure* (McGraw-Hill, New York, 1978).

25. N. C. Nigam and P. C. Jennings, *Bulletin of the Seismological Society of America* **59**, 909–922 (1969). Calculation of response spectra from strong earthquake records.

26. M. P. Singh and M. Ghafory-Ashtiany, *Earthquake Engineering and Structural Dynamics* **14**, 133–146 (1986). Modal time history analysis of non-classically damped structures for seismic motions.

27. B. L. Ly, *Journal of Sound and Vibration* **95**, 435–438 (1984). A computation technique for the response of linear systems.

28. E. A. Kraut, *Fundamentals of Mathematical Physics* (McGraw-Hill, New York, 1967).

29. R. E. Mickens, Difference equation models of differential equations having zero local truncation errors, in *Differential Equations*, I. W. Knowles and R. T. Lewis, editors (North-Holland, Amsterdam, 1984).

30. T. R. F. Nonweiler, *Computational Mathematics: An Introduction to Numerical Approximation* (Halsted Press, New York, 1984).

Chapter 6

Two First-Order, Coupled Ordinary Differential Equations

6.1 Introduction

In this chapter, we construct a class of finite difference schemes for two coupled first-order ordinary differential equations. These schemes have linear stability properties that are the same as the differential equation for all step-sizes. The differential equations considered are assumed to have a single (real) fixed-point. A major consequence of these schemes is the absence of elementary numerical instabilities. Briefly, numerical instabilities are solutions to the discrete finite difference equations that do not correspond to any of the solutions of the original differential equations [1, 2, 3]. (See Chapter 2.) Thus, the given finite difference equations are not able to model the correct mathematical (properties of the solutions to the differential equations [2]. In general, numerical instabilities will occur when the linear stability properties of the corresponding fixed-points of the differential and difference equations do not agree [4].

The work in this chapter extends the results of Mickens and Smith [3] and Mickens [4] to the case of the two coupled, first-order ordinary differential equations

$$\frac{dx}{dt} = F(x, y), \qquad (6.1.1)$$

$$\frac{dy}{dt} = G(x, y), \qquad (6.1.2)$$

that have only a single (real) fixed-point which can be chosen to be at the origin, i.e., $(\bar{x}, \bar{y}) = (0, 0)$, where $\bar{x}$ and $\bar{y}$ are simultaneous solutions to the equations

$$F(\bar{x}, \bar{y}) = 0 = G(\bar{x}, \bar{y}). \qquad (6.1.3)$$

After presenting certain background information in Section 6.2, we demonstrate how to explicitly construct discrete models for the linear parts of Eqs. (6.1.1) and (6.1.2) that have the correct linear stability properties for all values of the step-size. This result is then applied in Section 6.4 to the full nonlinear differential equation where we find that two major classes of discrete models emerge: the fully explicit and semi-explicit schemes. Section 6.5 gives a variety of examples of finite difference schemes constructed according to the rules of Section 6.4. Further generalizations and modifications of these rules are discussed for individual equations.

Finally, it should be mentioned that while the coupled differential equation systems considered in this chapter are a small subset of all possible such equations, they do describe many important dynamical systems. Examples include the van der Pol limit-cycle oscillator [5, 6]

$$\frac{dx}{dt} = y, \quad \frac{dy}{dt} = -x + \epsilon(1 - x^2)y, \tag{6.1.4}$$

where ϵ is a positive parameter; the Lewis oscillator [7, 8]

$$\frac{dx}{dt} = y, \quad \frac{dy}{dt} = -x + \epsilon(1 - |x|)y; \tag{6.1.5}$$

the Duffing oscillator [6]

$$\frac{dx}{dt} = y, \quad \frac{dy}{dt} = -x - \lambda x^3, \tag{6.1.6}$$

where λ is a parameter; and the modeling of batch fermentation processes [9]

$$\frac{dx}{dt} = -(A\alpha\beta)y - [(A\alpha)xy + (A\beta)y^2] - Axy^2, \tag{6.1.7}$$

$$\frac{dy}{dt} = -[(B\alpha\beta)x + (B\gamma\beta^2)y] + [(\alpha B)x^2] + (2B\beta\gamma)xy - (B\beta^2\epsilon)y^2]$$
$$+ [(B\gamma)x^2y - (2B\beta\epsilon)xy^2] - (B\epsilon)x^2y^2. \tag{6.1.8}$$

Other systems that can be modeled by two coupled, first-order ordinary differential equations arise in the biological sciences [10, 11], chemistry [12], and engineering [13, 14].

6.2   Background

In more detail, we assume that Eqs. (6.1.1) and (6.1.2) take the form

$$\frac{dx}{dt} = ax + by + f(x, y) \equiv F(x, y), \tag{6.2.1}$$

$$\frac{dy}{dt} = cx + dy + g(x, y) \equiv G(x, y), \tag{6.2.2}$$

where

$$ad - bc \neq 0, \tag{6.2.3}$$

and

$$f(x, y) = O(x^2 + y^2), \quad g(x, y) = O(x^2 + y^2). \tag{6.2.4}$$

We further assume that the only (real) solution to the equations

$$F(\bar{x}, \bar{y}) = 0, \quad G(\bar{x}, \bar{y}) = 0, \tag{6.2.5}$$

is

$$\bar{x} = 0, \quad \bar{y} = 0. \tag{6.2.6}$$

In general, for dynamical systems that model actual physical phenomena, the functions $F(x, y)$ and $G(x, y)$ are real analytic functions of x and y.

The concept of an exact finite difference scheme has already been defined and discussed in Section 3.2. The following theorem will be of value to the calculations given in Section 6.3.

Theorem. *Assume that the system of two coupled ordinary differential equations*

$$\frac{dX}{dt} = \Gamma(X), \quad X(t_0) \equiv X_0, \tag{6.2.7}$$

where

$$X = \begin{pmatrix} x \\ y \end{pmatrix}, \quad \Gamma(X) = \begin{pmatrix} F(x, y) \\ G(x, y) \end{pmatrix}, \tag{6.2.8}$$

has the solution

$$X(t) = \sigma(X_0, t_0, t). \tag{6.2.9}$$

Then Eq. (6.2.7) has the exact difference scheme

$$X_{k+1} = \sigma[X_k, hk, h(k + 1)], \tag{6.2.10}$$

where

$$X_k = X(hk). \tag{6.2.11}$$

We will now use this theorem "in reverse" to construct the exact finite difference scheme for the linear parts of Eqs. (6.2.1) and (6.2.2). See also Section 3.3.

6.3 Exact Scheme for Linear Ordinary Differential Equations

The terms in Eqs. (6.2.1) and (6.2.2) correspond to the following differential equation system

$$\frac{du}{dt} = au + bw, \tag{6.3.1}$$

$$\frac{dw}{dt} = cu + dw. \tag{6.3.2}$$

With the initial conditions

$$u_0 = u(t_0), \quad w_0 = w(t_0), \tag{6.3.3}$$

and the requirement

$$ad - bc \neq 0, \tag{6.3.4}$$

the general solution to Eqs. (6.3.1) and (6.3.2) are given by the relations [15]

$$\begin{aligned} u(t) = & -\left(\frac{b}{\lambda_1 - \lambda_2}\right)\left[\left(\frac{\lambda_2 - a}{b}\right)u_0 - w_0\right]e^{\lambda_1(t-t_0)} \\ & +\left(\frac{b}{\lambda_1 - \lambda_2}\right)\left[\left(\frac{\lambda_1 - a}{b}\right)u_0 - w_0\right]e^{\lambda_2(t-t_0)}, \end{aligned} \tag{6.3.5}$$

$$\begin{aligned} w(t) = & -\left(\frac{\lambda_1 - a}{\lambda_1 - \lambda_2}\right)\left[\left(\frac{\lambda_2 - a}{b}\right)u_0 - w_0\right]e^{\lambda_1(t-t_0)} \\ & +\left(\frac{\lambda_2 - a}{\lambda_1 - \lambda_2}\right)\left[\left(\frac{\lambda_1 - a}{b}\right)u_0 - w_0\right]e^{\lambda_2(t-t_0)}, \end{aligned} \tag{6.3.6}$$

where

$$2\lambda_{1,2} = (a+d) \pm \sqrt{(a+d)^2 - 4(ad - bc)}. \tag{6.3.7}$$

The exact difference scheme for Eqs. (6.3.1) and (6.3.2) is obtained by making the following substitutions in Eqs. (6.3.5) and (6.3.6):

$$\begin{cases} t_0 \to t_k = hk, \\ t \to t_{k+1} = h(k+1), \\ u_0 \to u_k, \\ u(t) \to u_{k+1}, \\ w_0 \to w_k, \\ w(t) \to w_{k+1}. \end{cases} \tag{6.3.8}$$

The results of these substitutions are

$$\frac{u_{k+1} - \psi u_k}{\phi} = au_k + bw_k, \tag{6.3.9}$$

$$\frac{w_{k+1} - \psi w_k}{\phi} = cu_k + dw_k, \tag{6.3.10}$$

where

$$\psi \equiv \psi(\lambda_1, \lambda_2, h) = \frac{\lambda_1 e^{\lambda_2 h} - \lambda_2 e^{\lambda_1 h}}{\lambda_1 - \lambda_2}, \tag{6.3.11}$$

$$\phi \equiv \phi(\lambda_1, \lambda_2, h) = \frac{e^{\lambda_1 h} - e^{\lambda_2 h}}{\lambda_1 - \lambda_2}. \tag{6.3.12}$$

The left-sides of Eqs. (6.3.9) and (6.3.10) are the discrete first-derivatives for Eqs. (6.3.1) and (6.3.2), i.e.,

$$\frac{du}{dt} \rightarrow \frac{u_{k+1} - \psi u_k}{\phi}, \tag{6.3.13}$$

$$\frac{dw}{dt} \rightarrow \frac{w_{k+1} - \phi w_k}{\phi}. \tag{6.3.14}$$

In the limit as

$$h \rightarrow 0, \quad k \rightarrow \infty, \quad hk = t = \text{fixed}, \tag{6.3.15}$$

these discrete derivatives reduce to the standard definition of the first-derivative since

$$\psi(\lambda_1, \lambda_2, h) = 1 + O(h^2), \tag{6.3.16}$$

$$\phi(\lambda_1, \lambda_2, h) = h + O(h^2). \tag{6.3.17}$$

However, note that for fixed, finite values of h, the nonstandard discrete derivatives given by Eqs. (6.3.13) and (6.3.14), do not agree with the definition of the discrete first-derivatives

$$\frac{du}{dt} \rightarrow \frac{u_{k+1} - u_k}{h}, \tag{6.3.18}$$

$$\frac{dw}{dt} \rightarrow \frac{w_{k+1} - w_k}{h}, \tag{6.3.19}$$

as given by the standard procedures [1, 3].

In summary, the finite difference model given by Eqs. (6.3.9) and (6.3.10) is the exact difference equation representation of Eqs. (6.3.1) and (6.3.2). As such, they satisfy, for any step-size h, the conditions

$$u_k = u(hk), \quad w_k = w(hk), \tag{6.3.20}$$

where $u(t)$ and $w(t)$ are the solutions to Eqs. (6.3.1) and (6.3.2), and u_k and w_k are the solutions to Eqs. (6.3.9) and (6.3.10).

6.4 Nonlinear Equations

The simplest finite difference scheme for the coupled, first-order nonlinear differential equations given by Eqs. (6.2.1) and (6.2.2) that has the correct linear stability properties for all values of the step-size is

$$\frac{x_{k+1} - \psi x_k}{\phi} = ax_k + by_k + f(x_k, y_k), \qquad (6.4.1)$$

$$\frac{y_{k+1} - \psi y_k}{\phi} = cx_k + dy_k + g(x_k, y_k), \qquad (6.4.2)$$

where ϕ and ψ are defined by Eqs. (6.3.11) and (6.3.12). This scheme evaluates the functions $f(x, y)$ and $g(x, y)$ at the same computational grid point, i.e., (x_k, y_k). Consequently, the discrete model of Eqs. (6.4.1) and (6.4.2) is explicit, i.e., both x_{k+1} and y_{k+1} are determined directly in terms of x_k and y_k.

A second possibility is the scheme

$$\frac{x_{k+1} - \psi x_k}{\phi} = ax_k + by_k + f(x_k, y_k), \qquad (6.4.3)$$

$$\frac{y_{k+1} - \psi y_k}{\phi} = cx_k + dy_k + g(x_{k+1}, y_k). \qquad (6.4.4)$$

Comparison with the previous discrete model shows that while Eq. (6.4.3) is the same as Eq. (6.4.1), Eqs. (6.4.4) and (6.4.2) differ. This difference occurs because in the $g(x, y)$ function the "x" variable is replaced by x_k in Eq. (6.4.2), but, by x_{k+1} in Eq. (6.4.4). This second scheme is a semi-explicit discrete model. By this we mean that for given values of (x_k, y_k), the value of x_{k+1} is first calculated from Eq. (6.4.3), then y_{k+1} is determined by Eq. (6.4.4). Thus, there is a definite order to how the calculation should be done.

In general, a variety of other discrete models can exist for Eqs. (6.2.1) and (6.2.2). In generic form, we indicate their structure by

$$\frac{x_{k+1} - \psi x_k}{\phi} = ax_k + by_k + f_k, \qquad (6.4.5)$$

$$\frac{y_{k+1} - \psi y_k}{\phi} = cx_k + dy_k + g_k, \qquad (6.4.6)$$

where f_k and g_k denote the particular discrete forms selected for $f(x, y)$ and $g(x, y)$. The important point is that all these schemes, including Eqs. (6.4.1) and (6.4.2), and Eqs. (6.4.3) and (6.4.4), have the property that their fixed-point at $(\bar{x}, \bar{y}) = (0, 0)$ has exactly the same linear stability behavior as the differential equation system for all step-sizes. Since the elementary numerical instabilities arise from a change in the linear stability properties of the fixed-points, it follows that these schemes will not have elementary numerical instabilities for any step-size.

In the section to follow, we will use the above results to construct non-standard finite difference models for a number of differential equations.

6.5 Examples

In this section discrete models of both conservative and limit-cycle oscillator systems that can be given as two coupled, first-order differential equations will be studied.

6.5.1 *Harmonic Oscillator*

The second-order harmonic oscillator differential equation is

$$\frac{d^2x}{dt^2} + x = 0. \tag{6.5.1}$$

In system form it becomes

$$\frac{dx}{dt} = y, \tag{6.5.2}$$

$$\frac{dy}{dt} = -x. \tag{6.5.3}$$

Comparing to Eqs. (6.2.1) and (6.2.2) gives

$$a = 0, \quad b = 1, \quad c = -1, \quad d = 0, \tag{6.5.4a}$$

$$f(x, y) = 0 \quad g(x, y) = 0. \tag{6.5.4b}$$

Substitution of these results into Eqs. (6.3.11) and (6.3.12) gives

$$\phi = \sin(h), \quad \psi = \cos(h). \tag{6.5.5}$$

Consequently, the exact difference scheme for the harmonic oscillator is

$$\frac{x_{k+1} - \cos(h)x_k}{\sin(h)} = y_k, \tag{6.5.6}$$

$$\frac{y_{k+1} - \cos(h)y_k}{\sin(h)} = -x_k. \tag{6.5.7}$$

A single second-order difference equation can be obtained by substituting the y_k of Eq. (6.5.6) into Eq. (6.5.7). This gives

$$\frac{[x_{k+2} - \cos(h)x_{k+1}] - \cos(h)[x_{k+1} - \cos(h)x_k]}{\sin^2(h)} = -x_k, \tag{6.5.8}$$

and

$$x_{k+2} - 2\cos(h)x_{k+1} + [\cos^2(h) + \sin^2(h)]x_k = 0. \tag{6.5.9}$$

But,

$$\cos^2(h) + \sin^2(h) = 1, \tag{6.5.10a}$$

$$2\cos(h) = 2 - 4\sin^2\left(\frac{h}{2}\right); \tag{6.5.10b}$$

therefore, Eq. (6.5.9) can be written as

$$(x_{k+1} - 2x_k + x_{k-1}) + \left[4\sin^2\left(\frac{h}{2}\right)\right]x_k = 0, \tag{6.5.11}$$

or, finally

$$\frac{x_{k+1} - 2x_k + x_{k-1}}{4\sin^2\left(\frac{h}{2}\right)} + x_k = 0. \tag{6.5.12}$$

This is precisely the form found earlier in Eq. (3.3.40).

6.5.2 *Damped Harmonic Oscillator*

The damped linear oscillator is described by the differential equation

$$\frac{d^2x}{dt^2} + 2\epsilon\frac{dx}{dt} + x = 0, \tag{6.5.13}$$

where ϵ is a positive constant. In system form this becomes

$$\frac{dx}{dt} = y, \tag{6.5.14}$$

$$\frac{dy}{dt} = -x - 2\epsilon y. \tag{6.5.15}$$

For this case, we have

$$a = 0, \quad b = 1, \quad c = -1, \quad d = -2\epsilon, \tag{6.5.16}$$

and the ψ and ϕ functions are

$$\psi(\epsilon, h) = \frac{\epsilon e^{-\epsilon h}}{\sqrt{1 - \epsilon^2}} + e^{-\epsilon h}\cos\left(\sqrt{1 - \epsilon^2}\right)h, \tag{6.5.17}$$

$$\phi(\epsilon, h) = \frac{e^{-\epsilon h}}{\sqrt{1 - \epsilon^2}} \sin\left(\sqrt{1 - \epsilon^2}\right) h. \qquad (6.5.18)$$

Thus, the exact finite difference scheme for the damped linear oscillator is

$$\frac{x_{k+1} - \psi x_k}{\phi} = y_k, \qquad (6.5.19)$$

$$\frac{y_{k+1} - \psi y_k}{\phi} = -x_k - 2\epsilon y_k. \qquad (6.5.20)$$

Using the expression for y_k, given by Eq. (6.5.19), a single second-order equation can be obtained for x_k; it is

$$\frac{x_{k+1} - 2\psi x_k + \psi^2 x_{k-1}}{\phi^2} + 2\epsilon\left(\frac{x_k - \psi x_{k-1}}{\phi}\right) + x_{k-1} = 0, \qquad (6.5.21)$$

where ψ and ϕ are given by Eqs. (6.5.17) and (6.5.18). An alternative form can be determined by transforming the various terms of Eq. (6.5.21). For example, multiplying by ϕ^2

$$[x_{k+1} - 2\psi x_k + \psi^2 x_{k-1}] + 2\epsilon\phi(x_k - \psi x_{k-1}) + \phi^2 x_{k-1} = 0, \qquad (6.5.22)$$

and using the relations

$$-2\psi x_k = -2x_k + 2(1 - \psi)x_k, \qquad (6.5.23)$$

$$(\phi^2 + \psi^2)x_{k-1} = x_{k-1} = x_{k-1} + (\phi^2 + \psi^2 - 1)x_{k-1}, \qquad (6.5.24)$$

gives

$$\begin{aligned}(x_{k+1} - 2x_k + x_{k-1}) + 2\epsilon\phi(x_k - \psi x_{k-1}) + 2(1 - \psi)x_k \\ + (\phi^2 + \psi^2 - 1)x_{k-1} = 0, \qquad (6.5.25)\end{aligned}$$

which on division by ϕ^2 is

$$\begin{aligned}\left[\frac{x_{k+1} - 2x_k + x_{k-1}}{\phi^2}\right] + 2\epsilon\left(\frac{x_k - \psi x_{k-1}}{\phi}\right) \\ + \left[\frac{2(1 - \psi)x_k + (\phi^2 + \psi^2 - 1)x_{k-1}}{\phi^2}\right]. \qquad (6.5.26)\end{aligned}$$

Comparison of either Eq. (6.5.21) or (6.5.26) with a standard finite difference scheme, such as

$$\frac{x_{k+1} - 2x_k + x_{k-1}}{h^2} + 2\epsilon\left(\frac{x_{k+1} - x_{k-1}}{2h}\right) + x_k = 0, \qquad (6.5.27)$$

demonstrates that they are clearly "nonstandard."

6.5.3 *Duffing Oscillator*

The nonlinear Duffing oscillator differential equation is

$$\frac{d^2x}{dt^2} + x + \beta x^3 = 0, \tag{6.5.28}$$

where β is a constant parameter. The first-order system equations are

$$\frac{dx}{dt} = y, \tag{6.5.29}$$

$$\frac{dy}{dt} = -x - \beta x^3. \tag{6.5.30}$$

For this case

$$a = 0, \quad b = 1, \quad c = -1, \quad d = 0, \tag{6.5.31}$$

$$f(x,y) = 0, \quad g(x,y) = -\beta x^3, \tag{6.5.32}$$

and

$$\psi \cos(h), \quad \phi = \sin(h). \tag{6.5.33}$$

The use of the explicit scheme of Eqs. (6.4.1) and (6.4.2) gives

$$\frac{x_{k+1} - \cos(h)x_k}{\sin(h)} = y_k, \tag{6.5.34}$$

$$\frac{y_{k+1} - \cos(h)y_k}{\sin(h)} = -x_k - \beta x_k^3. \tag{6.5.35}$$

The elimination of y_k gives

$$\frac{x_{k+1} - 2\cos(h)x_k + \cos^2(h)x_{k-1}}{\sin^2(h)} + x_{k-1} + \beta x_{k-1}^3 = 0, \tag{6.5.36}$$

which can be rewritten to the form

$$\frac{x_{k+1} - 2x_k + x_{k-1}}{4\sin^2\left(\frac{h}{2}\right)} + x_k + \beta \left[\frac{\sin^2(h)}{4\sin^2\left(\frac{h}{2}\right)} \right] x_{k-1}^3 = 0. \tag{6.5.37}$$

The corresponding semi-explicit scheme, based on Eqs. (6.4.3) and (6.4.4), is

$$\frac{x_{k+1} - \cos(h)x_k}{\sin(h)} = y_k, \tag{6.5.38}$$

$$\frac{y_{k+1} - \cos(h)y_k}{\sin(h)} = -x_k - \beta x_{k+1}^3. \tag{6.5.39}$$

Eliminating y_k and further manipulation of these results gives the expression

$$\frac{x_{k+1} - 2x_k + x_{k-1}}{4\sin^2\left(\frac{h}{2}\right)} + x_k + \beta \left[\frac{\sin^2(h)}{4\sin^2\left(\frac{h}{2}\right)}\right] x_k^3 = 0. \tag{6.5.40}$$

The question to be asked is which form, Eq. (6.5.37) or Eq. (6.5.40), should be used to calculate numerical solutions to the Duffing differential equation? It has been shown by Mickens [16] that the semi-explicit scheme is the one to use. (See, also, the arguments presented in Section 5.3.) The basic idea for this choice comes from the fact that the Duffing equation satisfies a conservation law. It follows that all the periodic solutions oscillate with constant amplitude. The semi-explicit scheme of Eq. (6.5.40) has this property, while the explicit scheme, given by Eq. (6.5.37), does not.

6.5.4 $\ddot{x} + x + \epsilon x^2 = 0$

This differential equation arises in the general theory of relativity [17]. Written as a system of first-order equations, it becomes

$$\frac{dx}{dt} = y, \tag{6.5.41}$$

$$\frac{dy}{dt} = -x - \epsilon x^2, \tag{6.5.42}$$

where ϵ is a constant parameter. Based on the result of Section 6.5.3, we will only consider the semi-explicit finite difference scheme, which for this problem is

$$\frac{x_{k+1} - \cos(h)x_k}{\sin(h)} = -y_k, \tag{6.5.43}$$

$$\frac{y_{k+1} - \cos(h)y_k}{\sin(h)} = -x_k - \epsilon x_{k+1}^2. \tag{6.5.44}$$

Eliminating y_k gives

$$\frac{x_{k+1} - 2x_k + x_{k-1}}{\sin^2(h)} + \frac{2[1 - \cos(h)]x_k}{\sin^2(h)} + \epsilon x_k^2 = 0. \tag{6.5.45}$$

Using the fact that

$$2[1 - \cos(h)] = 4\sin^2\left(\frac{h}{2}\right), \tag{6.5.46}$$

we finally obtain

$$\frac{x_{k+1} - 2x_k + x_{k-1}}{4\sin^2\left(\frac{h}{2}\right)} + x_k + \epsilon\left[\frac{\sin^2(h)}{4\sin^2\left(\frac{h}{2}\right)}\right]x_k^2 = 0, \qquad (6.5.47)$$

as a nonstandard discrete model for our original differential equation. Again, the arguments of Section 5.3 show that this finite difference scheme is conservative.

Note that another nonstandard discrete model is given by making the replacement

$$x^2 \to \frac{x_k(x_{k+1} + x_{k-1})}{2}, \qquad (6.5.48)$$

for the nonlinear x^2 term. The finite difference model in this case is given by the following expression

$$\frac{x_{k+1} - 2x_k + x_{k-1}}{4\sin^2\left(\frac{h}{2}\right)} + x_k + \epsilon\left[\frac{\sin^2(h)}{4\sin^2\left(\frac{h}{2}\right)}\right]\frac{x_k(x_{k+1} + x_{k-1})}{2} = 0. \quad (6.5.49)$$

This is also a conservative scheme that is semi-explicit since x_{k+1} can be calculated in terms of x_k and x_{k-1}. Based on the arguments of Section 5.3, we can conclude that the discrete model of Eq. (6.5.49) is to be preferred over that of Eq. (6.5.47).

6.5.5 *van der Pol Oscillator*

The van der Pol equation

$$\frac{d^2x}{dt^2} + x = \epsilon(1 - x^2)\frac{dx}{dt}, \qquad (6.5.50)$$

can be written in system form as

$$\frac{dx}{dt} = y, \qquad (6.5.51)$$

$$\frac{dx}{dt} = -x + \epsilon(1 - x^2)y, \qquad (6.5.52)$$

where ϵ is a positive parameter. Note that the linear terms of these equations are

$$\frac{du}{dt} = w, \qquad (6.5.53)$$

$$\frac{dw}{dt} = -u + \epsilon w, \qquad (6.5.54)$$

and correspond to an unstable "damped" linear oscillator. (See Eqs. (6.5.14) and (6.5.1).) For this case

$$a = 0, \quad b = 1, \quad c = -1, \quad d = \epsilon, \tag{6.5.55}$$

and the functions ϕ and ψ are given by the expressions

$$\psi(\epsilon, h) = -\frac{\epsilon e^{\epsilon h/2}}{2\sqrt{1 - \frac{\epsilon^2}{4}}} \sin\left(\sqrt{1 - \frac{\epsilon^2}{4}}\right) h$$

$$+ e^{\epsilon k/2} \cos\left(\sqrt{1 - \frac{\epsilon^2}{4}}\right) h, \tag{6.5.56}$$

$$\phi(\epsilon, h) = \frac{e^{\epsilon h/2}}{\sqrt{1 - \frac{\epsilon^2}{4}}} \sin\left(\sqrt{1 - \frac{\epsilon^2}{4}}\right) h. \tag{6.5.57}$$

Therefore, the semi-explicit scheme for the van der Pol differential equation is

$$\frac{x_{k+1} - \psi x_k}{\phi} = y_k, \tag{6.5.58}$$

$$\frac{y_{k+1} - \psi y_k}{\phi} = -x_k + \epsilon y_k - \epsilon x_{k+1}^2 y_k. \tag{6.5.59}$$

Eliminating y_k and rewriting the resulting expression gives

$$\frac{x_{k+1} - 2x_k + x_{k-1}}{\phi^2} + \frac{2(1 - \psi)x_k + (\psi^2 + \phi^2 - 1)x_{k-1}}{}$$

$$= \epsilon(1 - x_k^2) \left[\frac{x_k - \psi x_{k-1}}{\psi}\right]. \tag{6.5.60}$$

Another possibility for constructing a discrete model of the van der Pol equation is to consider the following set of linear terms

$$\frac{du}{dt} = w, \tag{6.5.61}$$

$$\frac{dw}{dt} = -u. \tag{6.5.62}$$

For this case, the linear term ϵy is incorporated into the function $g(x, y) = \epsilon(1 - x)y$. If we do this, the functions ψ and ϕ become

$$\psi = \cos(h), \quad \phi = \sin(h) \tag{6.5.63}$$

and, the semi-explicit scheme is

$$\frac{x_{k+1} - \cos(h)x_k}{\sin(h)} = y_k, \tag{6.5.64}$$

$$\frac{y_{k+1} - \cos(h)y_k}{\sin(h)} = -x_k + \epsilon(1 - x_{k+1}^2)y_k. \tag{6.5.65}$$

Eliminating y_k gives

$$\frac{x_{k+1} - 2x_k + x_{k-1}}{4\sin^2 2\left(\frac{h}{2}\right)} x_k = \epsilon \left[\frac{\sin(h)}{2\sin\left(\frac{h}{2}\right)}\right] (1 - x_k^2) \frac{x_k - \cos(h)x_{k-1}}{2\sin\left(\frac{h}{2}\right)}. \tag{6.5.66}$$

Two things should be noted about this last relation. First, it is similar to one of the discrete models investigated in Section 5.4. Second, this scheme does not have the correct linear stability properties: The van der Pol differential equation and the finite difference equation, given by Eq. (6.5.60), both have an unstable fixed-point $(\bar{x}, \bar{y}) = (0, 0)$; the scheme of Eq. (6.5.66) has neutral stability, i.e., the local stability properties of the harmonic oscillator. Thus, we conclude that Eq. (6.5.60) should be used as a discrete model for the van der Pol differential equation.

6.5.6   *Lewis Oscillator*

The differential equation for the nonlinear Lewis oscillator is [7]

$$\frac{d^2x}{dt^2} + x = \epsilon(1 - |x|)\frac{dy}{dt}, \tag{6.5.67}$$

where ϵ is a positive parameter. The corresponding system equations are

$$\frac{dx}{dt} = y, \tag{6.5.68}$$

$$\frac{dy}{dt} = -x + \epsilon y - \epsilon|x|y. \tag{6.5.69}$$

Since the linear terms of these equations are exactly the same as for the van der Pol differential equation, the functions ϕ and ψ for the Lewis oscillator are also given by Eqs. (6.5.56) and (6.5.57). Thus, the semi-explicit scheme is

$$\frac{x_{k+1} - \psi x_k}{\phi} = y_k, \tag{6.5.70}$$

$$\frac{y_{k+1} - \psi y_k}{\phi} = -x_k + \epsilon y_k - \epsilon|x_{k+1}|y_k, \tag{6.5.71}$$

or upon rewriting

$$\frac{x_{k+1} - 2x_k + x_{k-1}}{\phi^2} + \frac{2(1 - \psi)x_k + (\psi^2 + \phi^2 - 1)x_{k-1}}{}$$

$$= \epsilon(1 - |x_k|)\left[\frac{x_k - \psi x_{k-1}}{\phi}\right]. \tag{6.5.72}$$

6.5.7 *General Class of Nonlinear Oscillators*

Section 5.5 presented and discussed the construction of a nonstandard finite difference scheme for a general class of nonlinear oscillators for which the equation of motion is

$$\frac{d^2x}{dt^2} + xf(x^2)\frac{dx}{dt} + g(x^2)x = 0. \tag{6.5.73}$$

Written in system form, this equation becomes

$$\frac{dx}{dt} = y, \tag{6.5.74}$$

$$\frac{dy}{dt} = -x - g(x^2)x - f(x^2)y. \tag{6.5.75}$$

It is assumed that the functions $f(x^2)$ and $g(x^2)$ have the following properties:

$$f(x^2) = f_0 + f_1 x^2 + \bar{f}(x^2), \tag{6.5.76}$$

$$\bar{f}(x^2) = O(x^4), \tag{6.5.77}$$

$$g(0) = 0. \tag{6.5.78}$$

Consequently, the linear parts of Eqs. (6.5.7) and (6.5.75) are

$$\frac{du}{dt} = w, \tag{6.5.79}$$

$$\frac{dw}{dt} = -u - f_0 w, \tag{6.5.80}$$

and ψ and ϕ, in Eqs. (6.3.11) and (6.3.12), are to be calculated using

$$a = 0, \quad b = 1, \quad c = -1, \quad d = -f_0. \tag{6.5.81}$$

Therefore, the semi-explicit scheme for Eqs. (6.5.74) and (6.5.75) is

$$\frac{x_{k+1} - \psi x_k}{\phi} = y_k, \tag{6.5.82}$$

$$\frac{y_{k+1} - \psi y_k}{\phi} = -x_k - f_0 y_k - [f_1 x_{k+1}^2 + \bar{f}(x_{k+1}^2)]y_k - g(x_{k+1}^2)x_{k+1}. \tag{6.5.83}$$

Finally, eliminating y_k and rearranging the various terms gives

$$\left[\frac{x_{k+1} - 2x_k + x_{k-1}}{\phi^2}\right] + \left[\frac{2(1 - \psi)x_k + (\psi^2 + \phi^2 - 1)x_{k-1}}{\phi^2}\right]$$
$$+ f(x_k^2)\left[\frac{x_k - \psi x_{k-1}}{\phi}\right] + g(x_k^2)x_k = 0. \tag{6.5.84}$$

Combining this scheme with the nonlocal symmetric modeling of the $g(x^2)x$ term gives the following discrete representation

$$\left[\frac{x_{k+1} - 2x_k + x_{k-1}}{\phi^2}\right] + \frac{2(1 - \psi)x_k + (\psi^2 + \phi^2 - 1)x_{k-1}}{\phi^2}$$

$$+ f(x_k^2)\left[\frac{x_k - \psi x_{k-1}}{\phi}\right]$$

$$+ g(x_k^2)\left(\frac{x_{k+1} + x_{k-1}}{2}\right) = 0. \qquad (6.5.85)$$

Note that in contrast to Eq. (5.5.4), the discrete models, given in Eqs. (6.5.84) and (6.5.85), have the correct linear stability properties.

6.5.8   *Batch Fermentation Processes*

The differential equations are given by Eqs. (6.1.7) and (6.1.8). The linear terms correspond to the equations

$$\frac{du}{dt} = -(A\alpha\beta)w, \qquad (6.5.86)$$

$$\frac{dw}{dt} = (B\alpha\beta)u + (B\gamma\beta^2)w, \qquad (6.5.87)$$

where

$$a = 0, \quad b = -A\alpha\beta, \quad c = B\alpha\beta \quad d = B\gamma\beta^2. \qquad (6.5.88)$$

With these values, the functions ϕ and ψ can be determined using Eqs. (6.3.11) and (6.3.12). This gives the following semi-explicit finite difference scheme

$$\frac{x_{k+1} - \psi x_k}{\phi} = -(A\alpha\beta)y_k + [(A\alpha)x_k y_k + (A\beta)y_k^2] - Ax_k y_k^2, \qquad (6.5.89)$$

$$\frac{y_{k+1} - \psi y_k}{\phi} = [(B\alpha\beta)x_k + (B\gamma\beta^2)y_k]$$

$$+ [(\alpha B)x_{k+1}^2 + (2B\beta\gamma)x_{k+1}y_k - (B\beta^2\epsilon)y_k^2]$$

$$+ [(B\gamma)x_{k+1}^2 y_k - (2B\beta\epsilon)x_{k+1}y_k^2] - (B\epsilon)x_{k+1}^2 y_k^2. \qquad (6.5.90)$$

6.6 Summary

In this chapter, we have shown that it is possible to construct finite difference schemes for two coupled, first-order nonlinear differential equations, for which there is only a single (real) fixed-point, such that the difference equations have exactly the same linear stability properties as the differential equations for all finite values of the step-size. This result is very important since standard finite difference schemes, in general, do not have this property. In addition, this result implies that elementary numerical instabilities do not occur.

Based on the earlier work of Mickens [4, 16, 18], it was concluded that the semi-explicit procedure, given by Eqs. (6.4.3) and (6.4.4), is the proper discrete modeling technique to use. The semi-explicit scheme is an explicit method for which the variables are calculated in a definite order: first x_{k+1} is determined and then y_{k+1}.

As in the previous work of Mickens [2, 3, 4, 18], it was found that generalized representations of discrete derivatives appeared in the nonstandard finite difference schemes constructed in this chapter; see, for example, Eqs. (6.3.13) and (6.3.14). This feature is ubiquitous in the construction of nonstandard discrete models differential equations.

The semi-explicit scheme is not an exact finite difference discrete model. It is *a best finite difference scheme*. This means that it was constructed in such a way that a critical feature of its solution corresponded exactly to the related property of the original differential equation. In this case, the critical property was the nature of the stability for the fixed-point at $(\bar{x}, \bar{y}) = (0, 0)$.

References

1. F. B. Hildebrand, *Finite-Difference Equations and Simulations* (Prentice-Hall; Englewood Cliffs, NJ; 1968). Sections 2.6, 2.8 and 2.10.

2. R. E. Mickens, *Numerical Methods for Partial Differential Equations* **5**, 313–325 (1989). Exact solutions to a finite-difference model for a nonlinear reaction-advection equation: Implications for numerical analysis.

3. R. E. Mickens and A. Smith, *Journal of the Franklin Institute,* **327**, 143–149 (1990). Finite-difference models of ordinary differential equations: Influence of denominator functions.

4. R. E. Mickens, *Dynamic System and Applications* **1**, 329–340 (1992). Finite difference schemes having the correct linear stability properties for all finite step-sizes II.

5. B. van der Pol, *Philosophical Magazine* **43**, 177–193 (1922). On a type of oscillation hysteresis in a simple triode generator.

6. R. E. Mickens, *Nonlinear Oscillations* (Cambridge University Press, New York, 1981).

7. J. B. Lewis, *Transactions of the American Institute of Electrical Engineering, Part II* **72**, 449–453 (1953). The use of nonlinear feedback to improve the transient response of a servomechanism.

8. R. E. Mickens and I. Ramadhani, *Journal of Sound and Vibration* **154**, 190–193 (1992). Investigation of an anti-symmetric quadratic nonlinear oscillator.

9. Y. Lenbury, P. S. Crooke and R. D. Tanner, *BioSystems* **19**, 15–22 (1986). Relating damped oscillations to the sustained limit cycles describing real and ideal batch fermentation processes.

10. L. Edelstein-Keshet, *Mathematical Models in Biology* (Random House/Birkhauser, New York, 1988).

11. O. Sporns and F. F. Seelig, *BioSystems* **19**, 83–89 (1986). Oscillations in theoretical models of induction.

12. R. J. Field and M. Burger, editors, *Oscillating and Traveling Waves in Chemical Systems* (Wiley-Interscience, New York, 1985).

13. D. Potter, *Computational Physics* (Wiley-Interscience, New York, 1973).

14. J. M. T. Thompson and H. B. Stewart, *Nonlinear Dynamics and Chaos* (Wiley, New York, 1986).

15. S. L. Ross, *Differential Equations* (Xerox; Lexington, MA; 1974, 2nd edition). Chapter 7.

16. R. E. Mickens, *Journal of Sound and Vibration* **124**, 194–198 (1988). Properties of finite-difference models of nonlinear conservative oscillators.

17. W. Rindler, *Essential Relativity: Special, General and Cosmological* (Van Nostrand Reinhold, New York, 1969), section 7.5.

18. R. E. Mickens, Investigation of finite-difference models of the van der Pol equation, in *Differential Equations and Applications*, A. R. Aftabizadeh, editor (Ohio University Press; Columbus, OH; 1988), pp. 210–215.

Chapter 7

Partial Differential Equations

7.1 Introduction

Partial differential equations provide valuable mathematical models for dynamical systems that involve both space and time variables [1–8]. In this chapter, we study the construction of nonstandard finite difference schemes for a number of linear and nonlinear partial differential equations. In general, these equations are first-order in the time derivative and first- or second-order in the space derivatives. These equations include various one space dimension modifications of wave, diffusion and Burgers' partial differential equations. The nonlinearities considered are generally quadratic functions of the dependent variable and its derivatives. The use of only quadratic nonlinear terms follows from the result that for these expressions exact special solutions can often be found for the partial differential equation under study. These special solutions can then be used in the construction of nonstandard discrete models. However, it should be noted that exact finite difference schemes are not expected to exist for partial differential equations [9, 10]. This is a general consequence of the realization that for any arbitrarily specified partial differential equation, no precise definition of the general solution can be given [11].

For each partial differential equation considered, a comparison will be made to the standard finite difference schemes and how the solutions of the various nonstandard and nonstandard discrete models differ from each other. The results obtained in this chapter rely heavily on the concept of "best" finite difference scheme as introduced in Chapter 3. In summary, the best scheme is a discrete model of a differential equation that incorporates as many of the properties of the differential equation as possible into its mathematical structure. While best schemes correspond to nonstandard discrete models, they are not, in general, unique. Additional information or requirements are usually needed to obtain uniqueness.

Sections 7.2, 7.3 and 7.4 treat, respectively, the discrete modeling of wave, diffusion and Burgers' type partial differential equations. In Section 7.5, we summarize what has been found for the various finite difference schemes and carry out a general discussion on the problems of constructing better discrete models for partial differential equations.

7.2   Wave Equations

7.2.1 $u_t + u_x = 0$

The unidirectional wave equation

$$u_t + u_x = 0, \tag{7.2.1}$$

treated as an initial value problem, i.e.,

$$u(x, 0) = f(x) = \text{given}, \tag{7.2.2}$$

has the exact solution

$$u(x, t) = f(x - t). \tag{7.2.3}$$

This corresponds to a wave form traveling with unit velocity to the right.

A direct calculation [12] shows that the partial difference equation

$$u_m^{k+1} = u_{m-1}^k, \tag{7.2.4}$$

has the exact solution

$$u_m^k = h(m - k) \tag{7.2.5}$$

where $h(z)$ is an arbitrary function of z. For

$$\Delta x = \Delta t = h, \tag{7.2.6}$$

$$t_k = hk, \quad x_m = hm, \tag{7.2.7}$$

we have

$$u_m^k = h\left(\frac{x_m - t_k}{h}\right) = H(x_m - t_k). \tag{7.2.8}$$

Thus, it can be concluded that Eq. (7.2.4) is an exact finite difference model of the unidirectional wave equation given by Eq. (7.2.1), i.e.,

$$u_m^k = u(x_m, t_k), \tag{7.2.9}$$

where u_m^k is a solution of Eq. (7.2.4) and $u(x, t)$ is the corresponding solution of Eq. (7.2.1).

The above partial difference equation can be rewritten as

$$\frac{u_m^{k+1} - u_m^k}{\psi(\Delta t)} + \frac{u_m^k - u_{m-1}^k}{\psi(\Delta x)} = 0, \quad \Delta t = \Delta x, \tag{7.2.10}$$

where $\psi(z)$ has the property

$$\psi(z) = z + O(z^2). \tag{7.2.11}$$

The denominator function $\psi(z)$ is not determined by this analysis. Any $\psi(z)$ that satisfies the condition given in Eq. (7.2.11) will work. The simplest choice is $\psi(z) = z$. However, as Eq. (7.2.4) indicates, for the unidirectional wave equation, the particular choice is irrelevant since $\Delta t = \Delta x$ and the denominator functions drop out of the calculations. Note that the discrete time-derivative is forward-Euler, while the discrete space-derivative is a backward-Euler.

7.2.2 $u_t - u_x = 0$

There is a second linear unidirectional wave equation that describes a wave form traveling to the left with unit velocity. It is given by the equation

$$u_t - u_x = 0. \tag{7.2.12}$$

For the initial value problem

$$u(x, 0) = g(x) = \text{given}, \tag{7.2.13}$$

the exact solution is

$$u(x, t) = g(x + t). \tag{7.2.14}$$

The partial difference equation

$$u_m^{k+1} = u_{m+1}^k, \tag{7.2.15}$$

has the exact solution [12]

$$u_m^k = p(m + k). \tag{7.2.16}$$

Using the conditions of Eqs. (7.2.6) and (7.2.7), we have

$$u_m^k = P(x_m + t_k), \tag{7.2.17}$$

and, consequently, conclude that Eq. (7.2.15) is an exact finite difference model for Eq. (7.2.12).

Proceeding as in the last section, Eq. (7.2.15) can be rewritten to the form

$$\frac{u_m^{k+1} - u_m^k}{\psi(\Delta t)} - \frac{u_{m+1}^k - u_m^k}{\psi(\Delta x)} = 0, \tag{7.2.18}$$

where $\psi(z)$ has the property given by Eq. (7.2.11) and the condition

$$\Delta t = \Delta x, \tag{7.2.19}$$

must be satisfied. For this discrete model, both discrete first-derivatives are given by forward-Euler representations.

7.2.3 $u_{tt} - u_{xx} = 0$

The full or dual direction wave equation is

$$u_{tt} - u_{xx} = 0. \tag{7.2.20}$$

Its general solution is

$$u(x,t) = f(x - t) + g(x + t), \tag{7.2.21}$$

where $f(z)$ and $g(z)$ are arbitrary functions having second derivatives [4]. The exact finite difference equation for the wave equation is [4]

$$u_m^{k+1} + u_m^{k-1} = u_{m+1}^k + u_{m-1}^k. \tag{7.2.22}$$

To prove this, let us show that

$$w_m^k = F(m - k) + G(m + k) \tag{7.2.23}$$

is a solution to Eq. (7.2.22). Note that

$$u_m^{k+1} = F(m - k - 1) + G(m + k + 1), \tag{7.2.24}$$

$$u_m^{k-1} = F(m - k - 1) + G(m + k + 1), \tag{7.2.25}$$

and

$$u_{m+1}^k = F(m - k - 1) + G(m + k + 1), \tag{7.2.26}$$

$$u_{m-1}^k = F(m - k - 1) + G(m + k + 1). \tag{7.2.27}$$

Substitution of these expressions into, respectively, the left- and right-sides of Eq. (7.2.22) gives the desired result, namely Eq. (7.2.23) is the general solution.

Subtracting $2u_m^k$ from both sides of Eq. (7.2.22) and dividing by $\phi(h)$ where

$$\phi(x) = h^2 + O(h^4), \tag{7.2.28}$$

gives

$$\frac{u_m^{k+1} - 2u_m^k + u_m^{k-1}}{\phi(\Delta t)} = \frac{u_{m+1}^k - 2u_m^k + u_{m-1}^k}{\phi(\Delta x)}, \tag{7.2.29}$$

where the condition

$$\Delta t = \Delta x \tag{7.2.30}$$

is required. Note that the exact analytical expression for $\phi(h)$ is not needed since, with the condition of Eq. (7.2.30), the denominator functions drop out of the calculation.

7.2.4 $u_t + u_x = u(1 - u)$

The exact finite difference scheme for this nonlinear reaction-advection equation has already been given and discussed in Section 3.3 [9]. It is given by the expression

$$\frac{u_m^{k+1} - u_m^k}{\psi(\Delta t)} + \frac{u_m^k - u_{m-1}^k}{\psi(\Delta x)} = u_{m-1}^k(1 - u_m^{k+1}), \qquad (7.2.31)$$

where the denominator function is

$$\psi(h) = e^k - 1, \qquad (7.2.32)$$

and the following requirement must be satisfied

$$\Delta t = \Delta x. \qquad (7.2.33)$$

The following points should be observed:

(i) The discrete derivatives correspond exactly to those found previously for the unidirectional wave equation, the linear terms of this nonlinear differential equation.

(ii) The denominator function $\psi(h)$ has a specified form given by Eq. (7.2.32). This is a consequence of the nonlinearity of the partial differential equation.

(iii) There is a functional relation between the step-sizes, i.e., $\Delta t = \Delta h$.

(iv) The nonlinear u^2 term is modeled nonlocally on the discrete space-time grid, i.e.,

$$u^2 \to u_{m-1}^k u_m^{k+1}. \qquad (7.2.34)$$

(v) Standard finite difference schemes, such as

$$\frac{u_m^{k+1} - u_m^k}{\Delta t} + \frac{u_{m+1}^k - u_{m-1}^k}{\Delta x} = u_m^k(1 - u_m^{k+1}), \qquad (7.2.35)$$

do not have these features and, consequently, are expected to have numerical instabilities. (See references [8, 9, 12] and Section 2.5.)

7.2.5 $u_k + u_x = bu_{xx}$

The linear advection-diffusion equation

$$u_t + u_x = bu_{xx}, \quad b > 0, \qquad (7.2.36)$$

plays a very important role in the analysis of certain physical phenomena in fluid dynamics [2, 3]. Also, of equal importance is that this partial differential equation provides a good model for testing finite difference schemes constructed for the numerical integration of more complicated equations.

A vast literature exist on the detailed analysis of the stability properties for these finite difference schemes. These include the works of Peyret and Taylor [13], Chan [14], Strikewerda [15], and Bentley, Pinder and Herrera [16].

Two standard finite difference schemes that have been investigated in detail with regard to their stability properties are

$$\frac{u_m^{k+1} - u_m^k}{\Delta t} + \frac{u_{m+1}^k - u_{m-1}^k}{\Delta x} = b\left[\frac{u_{m+1}^k - 2u_m^k + u_{m-1}^k}{(\Delta x)^2}\right], \qquad (7.2.37)$$

and

$$\frac{u_m^{k+1} - u_m^k}{\Delta t} + \frac{u_m^k - u_{m-1}^k}{\Delta x} = b\left[\frac{u_{m+1}^k - 2u_m^k + u_{m-1}^k}{(\Delta x)^2}\right]. \qquad (7.2.38)$$

It should be clear that both of these schemes will give rise to numerical in stabilities. The first because it models the discrete space-derivative by a second-order central difference scheme and also because there is no relationship between the two step sizes, and the second because of the absence of a functional relation between Δt and Δx. (However, the requirements of stability do give a relationship between these two step-sizes [13–16].)

Our goal is to construct a conditionally stable explicit nonstandard finite difference model for the linear advection-diffusion equation [10]. (In brief, a conditional stable model is one for which the discrete-time dependency of the solution is bounded as $k \to \infty$. For details, see references [13, 14, 15, 17].) To begin, we note that Eq. (7.2.36) can be decomposed into the following two sub-equations

$$u_t + u_x = 0, \qquad (7.2.39)$$

$$u_x = bu_{xx}. \qquad (7.2.40)$$

Each of these differential equations has an exact discrete model. They are given, respectively, by the expression [18, 19]

$$\frac{u_m^{k+1} - u_m^k}{\psi(\Delta t)} + \frac{u_m^k - u_{m-1}^k}{\psi(\Delta x)} = 0, \qquad (7.2.41a)$$

$$\psi(h) = h + O(h^2), \quad \Delta x = \Delta t, \qquad (7.2.41b)$$

$$\frac{u_m - u_{m-1}}{\Delta x} = b\left[\frac{u_{m+1} - 2u_m + u_{m-1}}{b(e^{\Delta x/b} - 1)\Delta x}\right]. \qquad (7.2.42)$$

A finite difference model for Eq. (7.2.36) that combines both of these discrete subequations is

$$\frac{u_m^{k+1} - u_m^k}{\Delta t} + \frac{u_m^k - u_{m-1}^k}{\Delta x} = b\left[\frac{u_{m+1}^k - 2u_m^k}{b(e^{\Delta x/b} - 1)\Delta x)}\right]. \qquad (7.2.43)$$

For this equation, we do not know what the relation is between the step-sizes Δx and Δt. To find this relation, we proceed as follows. First, Eq. (7.2.43) can be solved for u_m^{k+1} to give

$$u_m^{k+1} = \beta u_{m+1}^k + (1 - \alpha - 2\beta)u_m^k + (\alpha + \beta)u_{m-1}^k, \qquad (7.2.44)$$

where

$$\alpha = \frac{\Delta t}{\Delta x}, \quad \beta = \frac{\alpha}{(e^{\Delta x/b} - 1)}. \qquad (7.2.45)$$

The conditional stability of the solutions to either Eq. (7.2.43) or (7.2.44) can be assured by the use of a result due to Forsythe and Wasow [17]. (See also reference [20].) They proved that if all the coefficients on the right-side of Eq. (7.2.44) are nonnegative, then the finite difference scheme is (conditionally) stable. This condition is related to the requirement that the solutions to Eq. (7.2.36) satisfy a min-max principle. However, in general, the solutions to Eq. (7.2.44) do not satisfy this principle. The imposing of this min-max principle on the solutions to Eq. (7.2.44) gives the conditional stability requirement.

The coefficients of the u_{m+1}^k and u_{m-1}^k terms are positive by definition of α and β, and the fact that $b > 0$. Hence, the conditional stability requirement is

$$1 - \alpha - 2\beta \geq 0, \qquad (7.2.46)$$

or

$$\Delta t \leq \Delta \left[\frac{e^{\Delta x/b} - 1}{e^{\Delta x/b} + 1} \right]. \qquad (7.2.47)$$

This inequality places a restriction on the time step-size once the space step-size is selected. By choosing the equality sign, we have a functional relation between Δx and Δt. Note that this relation

$$\Delta t = \Sigma(\Delta x, b) = \Delta x \left[\frac{e^{\Delta x/b} - 1}{e^{\Delta x/b} + 1} \right], \qquad (7.2.48)$$

also depends on the parameter b.

In summary, the linear advection-diffusion equation has a (nonstandard) best finite difference scheme given by Eq. (7.2.43) where the step-sizes are related by the result expressed in Eq. (7.2.48).

7.3   Diffusion Equations

The construction of better finite difference models for the linear diffusion equation

$$u_t = bu_{xx}, \quad b > 0, \tag{7.3.1}$$

has been studied since the beginning of modern numerical analysis [4, 6, 7, 15, 17, 20, 21, 22, 23]. The simplest scheme is the standard explicit forward-Euler which is given by the expression

$$\frac{u_m^{k+1} - u_m^k}{\Delta t} = b \left[\frac{u_{m+1}^k - 2u_m^k + u_{m-1}^k}{(\Delta x)^2} \right]. \tag{7.3.2}$$

The conditional stability requirement is [6]

$$\Delta \le \frac{(\Delta x)^2}{2b}. \tag{7.3.3}$$

This section will be devoted to an investigation of how nonstandard finite difference schemes can be constructed for diffusion type partial differential equations.

7.3.1 $u_t = au_{xx} + bu$

Consider the linear diffusion equation

$$u_t = au_{xx} + bu, \tag{7.3.4}$$

where a and b are constants with $a \ge 0$. The two sub-equations

$$\frac{du}{dt} = bu, \tag{7.3.5}$$

$$a\frac{d^2u}{dx^2} + bu = 0, \tag{7.3.6}$$

both have exact finite difference schemes. They are

$$\frac{u^{k+1} - u^k}{\left(\frac{e^{b\Delta t} - 1}{b} \right)} = bu^k, \tag{7.3.7}$$

$$a \left\{ \frac{u_{m+1} - 2u_m + u_{m-1}}{\left(\frac{4a}{b} \right) \sin^2 \left[\sqrt{\frac{b}{a}} \left(\frac{\Delta x}{2} \right) \right]} \right\} + bu_m = 0. \tag{7.3.8}$$

(Note that these results are correct for any value of the sign for b.) There are two ways of combining Eqs. (7.3.7) and (7.3.8) to form a discrete model for the full linear diffusion equation. The first is the explicit scheme

$$\frac{u_m^{k+1} - u_m^k}{\left(\frac{e^{b\Delta t} - 1}{b}\right)} = a \left\{ \frac{u_{m+1}^k - 2u_m^k + u_{m-1}^k}{\left(\frac{4a}{b}\right) \sin^2 \left[\sqrt{\frac{b}{a}} \left(\frac{\Delta x}{2}\right) \right]} \right\} + bu_m^k, \tag{7.3.9}$$

the second is the implicit scheme

$$\frac{u_m^{k+1} - u_m^k}{\left(\frac{e^{b\Delta t} - 1}{b}\right)} = a \left\{ \frac{u_{m+1}^{k+1} - 2u_m^{k+1} + u_{m-1}^{k+1}}{\left(\frac{4a}{b}\right) \sin^2 \left[\sqrt{\frac{b}{a}} \left(\frac{\Delta x}{2}\right) \right]} \right\} + bu_m^k. \tag{7.3.10}$$

Observe that in both schemes, the bu term is evaluated at the k-th discrete-time step rather than the $(k+1)$-th step. Also, both schemes reduce to the correct discrete model of the corresponding sub-equations.

Finite difference schemes for the simple diffusion equation are obtained by taking the limit $b \to 0$. Doing this for Eq. (7.3.9) gives the standard explicit model of Eq. (7.3.2), while Eq. (7.3.10) goes to the standard implicit form

$$\frac{u_m^{k+1} - u_m^k}{\Delta t} = a \left[\frac{u_{m+1}^{k+1} - 2u_m^{k+1} + u_{m-1}^{k+1}}{(\Delta x)^2} \right]. \tag{7.3.11}$$

7.3.2 $u_t = uu_{xx}$

The nonlinear diffusion equation

$$u_t - uu_{xx} \tag{7.3.12}$$

has the special rational solution

$$u(x, t) = \frac{-\left(\frac{\alpha}{2}\right) x^2 + \beta_2 x + \beta_2}{\alpha_1 + \alpha t}. \tag{7.3.13}$$

This can be shown by using the method of separation of variables and writing $u(x, t)$ as

$$u(x, t) = X(x)T(t), \tag{7.3.14}$$

and substituting this into Eq. (7.3.12) to obtain

$$\frac{1}{T^2} \frac{dT}{dt} = \frac{d^2 X}{dx^2} = -\alpha, \tag{7.3.15}$$

where α is the separation constant. The differential equations

$$\frac{dT}{dt} = -\alpha T^2, \quad \frac{d^2 X}{dx^2} = -\alpha, \tag{7.3.16}$$

have the respective solutions

$$T(t) = \frac{1}{\alpha_1 + \alpha t}, \tag{7.3.17}$$

$$X(x) = -\left(\frac{\alpha}{2}\right) x^2 + \beta_1 x + \beta_2, \tag{7.3.18}$$

where $(\alpha_1, \beta_1, \beta_2)$ are arbitrary integration constants.

Based on the nonstandard modeling rules, given in Section 3.4, the nonlinear term uu_{xx}, on the right-side of Eq. (7.3.12), must be modeled by a discrete form that is nonlocal in the time variable. The simplest two choices are

$$uu_{xx} \rightarrow \begin{cases} u_m^{k+1} \left[\frac{u_{m+1}^k - 2u_m^k + u_{m-1}^k}{\phi(\Delta x)} \right], \\ u_m^k \left[\frac{u_{m+1}^1 - 2u_m^{k+1} + u_{m-1}^{k+1}}{\phi(\Delta x)} \right], \end{cases} \tag{7.3.19}$$

where the denominator function ϕ has the property

$$\phi(h) = h^2 + O(h^4). \tag{7.3.20}$$

These lead to the following two finite difference schemes

$$\frac{u_m^{k+1} - u_m^k}{\psi(\Delta t)} = u_m^{k+1} \left[\frac{u_{m+1}^k - 2u_m^k + u_{m-1}^k}{\phi(\Delta x)} \right], \tag{7.3.21}$$

$$\frac{u_m^{k+1} - u_m^k}{\psi(\Delta t)} = u_m^k \left[\frac{u_{m+1}^{k+1} - 2u_m^{k+1} + u_{m-1}^{k+1}}{\phi(\Delta x)} \right], \tag{7.3.22}$$

where

$$\psi(h) = h + O(h^2). \tag{7.3.23}$$

These models are, respectively, explicit and implicit finite difference schemes for Eq. (7.3.12).

Consider first Eq. (7.3.21). A special exact solution can be found by the method of separation of variables [12], i.e., take u_m^k to be

$$u_m^k = C^k D_m. \tag{7.3.24}$$

where C^k is a function only of the discrete-time k and D_m depends only on the discrete-space variable m. Substitution of Eq. (7.3.24) into Eq. (7.3.21) gives

$$\frac{(C^{k+1} - C^k)D_m}{\psi} = C^{k+1} C^k D_m \left[\frac{D_{m+1} - 2D_m + D_{m-1}}{\phi} \right] \tag{7.3.25}$$

and

$$\frac{C^{k+1} - C^k}{\psi C^{k+1} C^k} = \frac{D_{m+1} - 2D_m + D_{m-1}}{\phi} = -\alpha, \qquad (7.3.26)$$

where α is the separation constant. The two ordinary difference equations

$$C^{k+1} - C^k = -\alpha\psi C^{k+1} C^k, \qquad (7.3.27)$$

$$D_{m+1} - 2D_m + D_{m-1} = -\alpha\phi, \qquad (7.3.28)$$

have the respective solutions [12]

$$C^k = \frac{1}{A_1 + \alpha\psi k}, \qquad (7.3.29)$$

$$D_m = -\left(\frac{\alpha}{2}\right)\psi m^2 + B_1 m + B_2, \qquad (7.3.30)$$

where (A_1, B_1, B_2) are arbitrary constants. Comparison of Eqs. (7.3.17) and (7.3.29), and (7.3.18) and (7.3.30) shows that if we require for the special rational solution

$$u_m^k = u(x_m, t_k), \qquad (7.3.31)$$

then we must have

$$\psi(\Delta t) = \Delta t, \quad \phi(\Delta x) = (\Delta x)^2, \qquad (7.3.32)$$

and

$$A_1 = \alpha_1, \quad B_1 = (\Delta x)\beta_1, \quad B_2 = \beta_2. \qquad (7.3.33)$$

Thus, we conclude that the explicit finite difference scheme

$$\frac{u_m^{k+1} - u_m^k}{\Delta t} = u_m^{k+1}\left[\frac{u_{m+1}^k - 2u_m^k + u_{m-1}^k}{(\Delta x)^2}\right], \qquad (7.3.34)$$

has exactly the same rational solution as the original nonlinear diffusion equation given by Eq. (7.3.13). The same set of steps shows that the implicit scheme of Eq. (7.3.22) also has this property provided that the denominator functions $\psi(\Delta t)$ and $\phi(\Delta x)$ are those of Eq. (7.3.32). This implicit scheme is

$$\frac{u_m^{k+1} - u_m^k}{\Delta t} = u_m^k\left[\frac{u_{m+1}^{k+1} - 2u_m^{k+1} + u_{m-1}^{k+1}}{(\Delta x)^2}\right], \qquad (7.3.35)$$

With no other restrictions, we cannot choose between these discrete models. However, experience with the standard techniques of numerical analysis indicates that the implicit scheme should provide "better" numerical results [6, 7].

It should be clear that a standard finite difference model, such as

$$\frac{u_m^{k+1} - u_m^k}{\Delta t} = u_m^k \left[\frac{u_{m+1}^k - 2u_m^k + u_{m-1}^k}{(\Delta x)^2} \right], \qquad (7.3.36)$$

cannot have the exact rational solution of Eqs. (7.3.29) and (7.3.30). The separation of variables form $u_m^k = C^k D_m$ gives for C^k and D_m the equations

$$\frac{(C^{k+1} - C^k)D_m}{\Delta t} = (C^k)D_m \left[\frac{D_{m+1} - 2D_m + D_{m-1}}{(\Delta x)^2} \right] \qquad (7.3.37)$$

and

$$C^{k+1} - C^k = -\alpha(\Delta t)(C^k)^2, \qquad (7.3.38)$$

$$D_{m+1} - 2D_m + D_{m-1} = -\alpha(\Delta x)^2, \qquad (7.3.39)$$

where α is the separation constant. The equation for C^k is the logistic difference equation and has a variety of solution behaviors, none of which are given by Eq. (7.3.29). Hence, the discrete model of Eq. (7.3.36) will have numerical instabilities.

Finally, it should be observed that at this stage of the investigation no relationship exists between the two step-sizes, Δx and Δt.

7.3.3 $u_t = uu_{xx} + \lambda u(1 - u)$

The nonlinear diffusion equation [24]

$$u_t = uu_{xx} + \lambda u(1 - u) \qquad (7.3.40)$$

can be decomposed in to three special limiting cases. They are (i) the space-independent equation

$$u_t = \lambda u(1 - u), \qquad (7.3.41)$$

(ii) the time-independent equation

$$u_{xx} + \lambda(1 - u) = 0, \qquad (7.3.42)$$

and (iii) the $\lambda = 0$ equation

$$u_t = u_{xx}. \qquad (7.3.43)$$

The first two equations are ordinary differential equations for which exact finite difference schemes exist. They are

$$\frac{u^{k+1} - u^k}{\left(\frac{e^{\lambda \Delta t} - 1}{\lambda} \right)} = \lambda u^k (1 - u^{k+1}), \qquad (7.3.44)$$

$$\frac{u_{m+1} - 2u_m + u_{m-1}}{\left(\frac{4}{\lambda}\right)\sinh^2\left(\frac{\sqrt{\lambda}\Delta t}{2}\right)} + \lambda(1 - u_m) = 0. \tag{7.3.45}$$

In the previous section we derived best difference schemes for the $\lambda = 0$ equation, namely, Eqs. (7.3.34) and (7.3.35).

We must now combine Eqs. (7.3.44), (7.3.45) and either Eq. (7.3.34) or (7.3.35) to obtain a discrete model for Eq. (7.3.40). For an explicit scheme there is only one way to do this and the proper scheme is

$$\frac{u_m^{k+1} - u_m^k}{\left(\frac{e^{\lambda\Delta t}-1}{\lambda}\right)} = u_m^{k+1}\left[\frac{u_{m+1}^k - 2u_m^k + u_{m-1}^k}{\left(\frac{4}{\lambda}\right)\sinh^2\left(\frac{\sqrt{\lambda}\Delta t}{2}\right)}\right] + \lambda u_m^k(1 - u_m^{k+1}). \tag{7.3.46}$$

The corresponding implicit scheme is

$$\frac{u_m^{k+1} - u_m^k}{\left(\frac{e^{\lambda\Delta t}-1}{\lambda}\right)} = u_m^k\left[\frac{u_{m+1}^{k+1} - 2u_m^{k+1} + u_{m-1}^{k+1}}{\left(\frac{4}{\lambda}\right)\sinh^2\left(\frac{\sqrt{\lambda}\Delta x}{2}\right)}\right] + \lambda u_m^k(1 - u_m^{k+1}). \tag{7.3.47}$$

In contrast to a standard finite difference model of Eq. (7.3.40), i.e.,

$$\frac{u_m^{k+1} - u_m^k}{\Delta t} = u_m^k\left[\frac{u_{m+1}^k - 2u_m^k + u^k - [m-1]}{(\Delta x)^2}\right] + \lambda u_m^k(1 - u_m^{k+1}), \tag{7.3.48}$$

for which we expect a variety of numerical instabilities to occur, the schemes of Eqs. (7.3.46) and (7.3.47) have the following properties:

(i) They have the correct discrete forms for the three special limiting differential equations.

(ii) Nonstandard forms for the discrete derivatives occur.

(iii) The nonlinear terms are modeled by nonlocal discrete expressions on the discrete-space and -time lattice.

7.3.4 $u_t = u_{xx} + \lambda u(1 - u)$

A famous differential equation that was originally used to model mutant-gene propagation is the Fisher equation [25]

$$u_t = u_{xx} + \lambda u(1 - u), \quad \lambda > 0. \tag{7.3.49}$$

Discrete versions of this equation have been used to investigate numerical instabilities in finite difference schemes [26, 27].

The Fisher equation has the two sub-equations that are ordinary differential equations. They are

$$u_t = \lambda u(1 - u), \tag{7.3.50}$$

$$u_{xx} + \lambda u(1 - u) = 0. \tag{7.3.51}$$

The exact finite difference scheme for the first of these equations is given by Eq. (7.3.44). The second differential equation corresponds to a nonlinear conservative oscillator [28]. We now construct a best finite difference model for it that satisfies an energy conservation principle.

A first integral for Eq. (7.3.51) is [28]

$$\left(\frac{1}{2}\right) u_x^2 + \lambda \left[\frac{u^2}{2} - \frac{u^3}{3}\right] = E = \text{constant.} \tag{7.3.52}$$

The application of standard modeling rules to Eq. (7.3.52) gives

$$\left(\frac{1}{2}\right) \left[\frac{u_m - u_{m-1}}{h}\right]^2 + \lambda \left[\frac{u_m^2}{2} - \frac{u_m^3}{3}\right] = E. \tag{7.3.53}$$

However, this expression does not correspond to a nonlinear oscillator with a conservation law since it is not invariant under the transformation

$$u_m \leftrightarrow u_{m-1}. \tag{7.3.54}$$

(See the discussion presented in Section 5.3.) A nonstandard scheme that does have this property is [29]

$$\left(\frac{1}{2}\right) \left[\frac{u_m - u_{m-1}}{\psi(\Delta t)}\right]^2 + \lambda \left\{\frac{u_m u_{m-1}}{2} - \left(\frac{u_m^2 u_{m-1} + u_m u_{m-1}^2}{6}\right)\right\} = E, \tag{7.3.55}$$

where

$$\psi(h) = h + O(h^2). \tag{7.3.56}$$

Applying the "difference operator," defined as

$$\Delta f_m \equiv f_{m+1} - f_m, \tag{7.3.57}$$

to Eq. (7.3.55) gives the discrete equation of motion

$$\frac{u_{m+1} - 2u_m + u_{m-1}}{[\psi(\Delta t)]^2} + \lambda u_m - \lambda \left(\frac{u_{m+1} + u_m + u_{m-1}}{3}\right) u_m = 0, \tag{7.3.58}$$

where the following results have been used:

$$\begin{aligned}
\Delta(u_m - u_{m-1})^2 &= \Delta(u_m^2 - 2u_m u_{m-1} + u_{m-1}^2) \\
&= (u_{m+1}^2 - u_m^2) - 2(u_{m+1} u_m - u_m u_{m-1}) + (u_m^2 - u_{m-1}^2) \\
&= (u_{m+1}^2 - u_{m-1}^2) - 2u_m(u_{m+1} - u_{m-1}) \\
&= (u_{m+1} - 2u_m + u_{m-1})(u_{m+1} - u_{m-1}), \tag{7.3.59}
\end{aligned}$$

$$\Delta u_m u_{m-1} = u_m(u_{m+1} - u_{m-1}), \tag{7.3.60}$$

$$\Delta(u_m^2 u_{m-1} + u_m u_{m+1}^2) = u_m(u_{m+1} + u_m + u_{m-1})(u_{m+1} - u_{m-1}). \quad (7.3.61)$$

The two discrete sub-equations, Eqs. (7.3.44) and (7.3.58), must now be combined. The only way to do this, to obtain an explicit scheme, is to use the representation

$$\frac{u_m^{k+1} - u_m^k}{\left(\frac{e^{\lambda \Delta t} - 1}{\lambda}\right)} = \frac{u_{m+1}^k - 2u_m^k + u_{m-1}^k}{\left(\frac{4}{\lambda}\right) \sin^2\left(\frac{\sqrt{\lambda} \Delta x}{2}\right)} + \lambda u_m^k$$
$$- \lambda \left(\frac{u_{m+1}^k + u_m^k + u_{m-1}^k}{3}\right) u_m^{k+1}. \quad (7.3.62)$$

Corresponding implicit schemes are

$$\frac{u_m^{k+1} - u_m^k}{\left(\frac{e^{\lambda \Delta t} - 1}{\lambda}\right)} = \frac{u_{m+1}^{k+1} - 2u_m^{k+1} + u_{m-1}^{k+1}}{\left(\frac{4}{\lambda}\right) \sin^2\left(\frac{\sqrt{\lambda} \Delta x}{2}\right)} + \lambda u_m^k$$
$$- \lambda \left(\frac{u_{m+1}^k + u_m^k + u_{m-1}^k}{3}\right) u_m^{k+1}, \quad (7.3.63)$$

and

$$\frac{u_m^{k+1} - u_m^k}{\left(\frac{e^{\lambda \Delta t} - 1}{\lambda}\right)} = \frac{u_{m+1}^{k+1} - 2u_m^{k+1} + u_{m-1}^{k+1}}{\left(\frac{4}{\lambda}\right) \sin^2\left(\frac{\sqrt{\lambda} \Delta x}{2}\right)} + \lambda u_m^k$$
$$- \lambda \left(\frac{u_{m+1}^{k+1} + u_m^{k+1} + u_{m-1}^{k+1}}{3}\right) u_m^k. \quad (7.3.64)$$

Note that the λu term must be evaluated at the m-th discrete-space step and the k-th discrete-time step.

None of the above schemes are even distantly related to the following scheme often used for calculations [26]

$$\frac{u_m^{k+1} - u_m^k}{\Delta t} = \frac{u_{m+1}^k - 2u_m^k + u_{m-1}^k}{(\Delta x)^2} + \lambda u_m^k(1 - u_m^k). \quad (7.3.65)$$

7.4 Burgers' Type Equations

The Burgers' partial differential equation [2, 3, 30]

$$u_t + uu_x = \nu u_{xx}, \quad \nu = \text{constant} \quad (7.4.1)$$

is a simplification of the Navier-Stokes equations. It is also the governing equation for a variety of one-dimensional flow problems, including, for example, weak shock propagation, compressible turbulence, and continuum traffic simulations [2]. In this section, we present a number of nonstandard discrete models for Burgers' type partial differential equations, i.e., equations having the form

$$u_t + uu_x = h_1(u)u_{xx} + h_2(u). \quad (7.4.2)$$

7.4.1 $u_t + uu_x = 0$

The diffusion-free Burgers' equation is

$$u_t + uu_x = 0. \tag{7.4.3}$$

With the initial condition

$$u(x,0) = f(x), \tag{7.4.4}$$

the exact solution is [2]

$$u(x,t) = f[x - u(x,t)t]. \tag{7.4.5}$$

The Burgers' equation has an exact rational solution that can be found by the method of separation of variables. If we write

$$u(x,t) = X(x)T(t), \tag{7.4.6}$$

then $X(x)$ and $T(t)$ satisfy the ordinary differential equations

$$\frac{dT}{dt} = \alpha T^2, \tag{7.4.7}$$

$$\frac{dX}{dx} = -\alpha, \tag{7.4.8}$$

where α is the separation constant. Solving these equations gives

$$T(t) = \frac{1}{A_1 - \alpha t}, \tag{7.4.9}$$

$$X(t) = A_2 - \alpha x, \tag{7.4.10}$$

where A_1 and A_2 are arbitrary constants of integration. Consequently, a special solution of the Burgers' equation is

$$u(x,t) = \frac{A_2 - \alpha x}{A_1 - \alpha t}. \tag{7.4.11}$$

We now require that our finite difference models for Eq. (7.4.3) have the discrete form of Eq. (7.4.11) as a special exact solution. In addition, we will also impose the condition that the nonlinear term, uu_x, be modeled nonlocally on the discrete-space and -time lattice.

The following two schemes have these properties [19]:

$$\frac{u_m^{k+1} - u_m^k}{\Delta t} + u_m^{k+1}\left(\frac{u_m^k - u_{m-1}^k}{\Delta x}\right) = 0, \tag{7.4.12}$$

$$\frac{u_m^{k+1} - u_m^k}{\Delta t} + u_m^k\left(\frac{u_m^{k+1} - u_{m-1}^{k+1}}{\Delta x}\right) = 0. \tag{7.4.13}$$

Note that applying the method of separation of variables [12] gives for both equations the expressions

$$D_{m+1} - D_m = -\alpha(\Delta x), \tag{7.4.14}$$

$$C^{k+1} - C^k = \alpha(\Delta t)C^{k+1}C^k, \tag{7.4.15}$$

where

$$u_m^k = C^k D_m, \tag{7.4.16}$$

and α is the separation constant. The solutions to these first-order difference equations can be put in the forms

$$D_m = A_2 - \alpha x_m, \tag{7.4.17}$$

$$C^k = \frac{1}{A_1 - \alpha t_k}, \tag{7.4.18}$$

where A_1 and A_2 are arbitrary constants. Therefore,

$$u_m^k = C^k D_m = \frac{A_2 - \alpha x_k}{A_1 - \alpha t_k}, \tag{7.4.19}$$

and, as stated above, Eqs. (7.4.12) and (7.4.13) both have the same special solutions as the diffusion-free Burgers' equation. Observe that the finite difference schemes of Eqs. (7.4.12) and (7.4.13) are, respectively, explicit and implicit.

7.4.2 $u_t + uu_x = \lambda u(1 - u)$

The following modified Burgers' equation

$$u_t + uu_x = \lambda u(1 - u), \tag{7.4.20}$$

where λ is a positive constant, has the three sub-equations

$$u_t = \lambda u(1 - u), \tag{7.4.21}$$

$$u_x = \lambda(1 - u), \tag{7.4.22}$$

$$u_t + uu_x = 0. \tag{7.4.23}$$

The first two equations are ordinary differential equations, while the third is the partial differential equation discussed in the previous section. The exact difference schemes, respectively, for Eqs. (7.4.21) and (7.4.22) are

$$\frac{u^{k+1} - u^k}{\left(\frac{e^{\lambda \Delta t}-1}{\lambda}\right)} = \lambda u^k(1 - u^{k+1}), \tag{7.4.24}$$

$$\frac{u_{m+1} - u_m}{\left(\frac{1-e^{\lambda\Delta x}-1}{\lambda}\right)} = \lambda(1 - u_m), \tag{7.4.25}$$

while best difference schemes for Eq. (7.4.23) are given by Eqs. (7.4.12) and (7.4.13).

We now require that any discrete model of Eq. (7.4.20) reduces, in the appropriate limit, to the finite difference results given by Eqs. (7.4.24), (7.4.25) and either Eq. (7.4.12) or (7.4.13). There is only one way that this can be done and this scheme is given by the expression

$$\frac{u_m^{k+1} - u_m^k}{\phi(\Delta t)} + u_m^k \left[\frac{u_m^{k+1} - u_{m-1}^{k+1}}{\psi(\Delta x)}\right] = \lambda u_m^k (1 - u_{m-1}^{k+1}), \tag{7.4.26}$$

where the denominator functions are

$$\phi(\Delta t) = \frac{e^{\lambda\Delta t} - 1}{\lambda}, \quad \psi(\Delta x) = \frac{1 - e^{\lambda\Delta x}}{\lambda}. \tag{7.4.27}$$

This finite difference scheme has the following properties:

(i) The discrete model is implicit.

(ii) The denominator functions are not of the simple forms $\phi(\Delta t) = \Delta t$ and $\psi(\Delta x) = \Delta x$.

(iii) The nonlinear terms are modeled nonlocally on the discrete-space and -time lattice.

(iv) The discrete space-derivative is a backward-Euler type scheme.

7.4.3 $u_t + uu_x = uu_{xx}$

Consider the following modified Burgers' equation [32]

$$u_t + uu_x = uu_{xx}. \tag{7.4.28}$$

This nonlinear partial differential equation does not have a known exact general solution that can be written in terms of a finite number of elementary functions. However, a special solution can be found by use of the method of separation of variables. Assuming for $u(x,t)$ the form

$$u(x,t) = X(x)T(t), \tag{7.4.29}$$

we find that

$$u(x,t) = \frac{A + Be^x + Cx}{Ct + D}, \tag{7.4.30}$$

where (A, B, C, D) are arbitrary constants. Therefore, we require that our finite difference schemes also have the discrete version of Eq. (7.4.30) as a special solution.

The application of the standard rules to Eq. (7.4.28) gives, for example, the form

$$\frac{u_m^{k+1} - u_m^k}{\Delta t} + u_m^k \left(\frac{u_{m+1}^k - u_{m-1}^k}{2\Delta x} \right) = u_m^k \left[\frac{u_{m+1}^k - 2u_m^k + u_{m-1}^k}{(\Delta x)^2} \right].$$
(7.4.31)

The method of variables can be applied to this difference equation. Writing u_m^k, as

$$u_m^k = X_m T^k,$$
(7.4.32)

we find that X_m and T^k satisfy the equations

$$\frac{X_{m+1} - 2X_m + X_{m-1}}{(\Delta x)^2} - \left[\frac{X_{m+1} - X_{m-1}}{2\Delta x} \right] = -C_1,$$
(7.4.33)

$$\frac{T^{k+1} - T^k}{\Delta t} = -C(\Delta t)(T^k)^2,$$
(7.4.34)

where C is the separation constant. The solutions to these equations do not correspond to the discrete versions of $X(x)$ and $T(t)$ as given in Eq. (7.4.30).

Now consider the following nonstandard model for Eq. (7.4.28) [32]:

$$\frac{u_m^{k+1} - u_m^k}{\Delta t} + u_m^{k+1} \left(\frac{u_m^k - u_{m-1}^k}{2\Delta x} \right) = u_m^{k+1} \left[\frac{u_{m+1}^k - 2u_m^k + u_{m-1}^k}{(e^{\Delta x} - 1)\Delta x} \right].$$
(7.4.35)

This result is obtained by combining the finite difference schemes for the three subequations

$$u_t + uu_x = 0,$$
(7.4.36)

$$u_t = uu_{xx},$$
(7.4.37)

$$u_x = u_{xx}.$$
(7.4.38)

Best schemes for Eqs. (7.4.36) and (7.4.37) have already been given, see Eqs. (7.4.12) and (7.3.34). The exact scheme for Eq. (7.4.38) is [19]

$$\frac{u_m - u_{m-1}}{\Delta x} = \frac{u_{m+1} - 2u_m + u_{m-1}}{(e^{\Delta x} - 1)\Delta x}.$$
(7.4.39)

A detailed examination of Eq. (7.4.35) leads to the following conclusions:

(i) This nonstandard scheme has an exact solution that can be found from the method of separation of variables. It is given by the discrete form of Eq. (7.4.30), namely,

$$u_m^k = \frac{A + Be^{x_m} + Cx_m}{Ct_k + D}.$$
(7.4.40)

See reference [32].

(ii) All the nonlinear terms of Eq. (7.4.28) are modeled nonlocally.

(iii) The denominator function for the discrete second-derivative has a nonstandard form.

(iv) The first-derivative is given by a backward-Euler type expression.

(v) The finite difference scheme is explicit.

(vi) A fully implicit scheme is given by the expression

$$\frac{u_m^{k+1} - u_m^k}{\Delta t} + u_m^k \left(\frac{u_m^{k+1} - u_{m-1}^{k+1}}{\Delta x} \right) = u_m^k \left[\frac{u_{m+1}^{k+1} - 2u_m^{k+1} + u_{m-1}^{k+1}}{(e^{\Delta x} - 1)\Delta x} \right].$$

$$(7.4.41)$$

7.5   Discussion

In general, we do not expect exact finite difference schemes to exist for arbitrary differential equations [18,19]. However, in practical applications, best schemes can be found and they should provide better discrete models than those obtained from use of standard methods [8, 10, 24, 29, 31, 32]. However, the work of this chapter shows that for a given partial differential equation, more than one best scheme may exist. This non-uniqueness usually appears in the form of the existence of both explicit and implicit best schemes for the equation of interest. Resolution of this problem can only come from imposing additional requirements on the finite difference schemes.

The "toy" partial differential equations considered in this chapter were investigated because they have special solutions that can be found and/or they have subequations such that these equations can be solved exactly or have special solutions that can be discovered. Our basic procedure for constructing best finite difference schemes consisted of imposing one or both of the following requirements on the discrete model:

(i) Special solutions of the differential equation should also be special solutions of the finite difference equation.

(ii) Corresponding sub-equations of the differential and finite difference equations should have (essentially) the same mathematical properties. In particular, this means that in the proper limits, the sub-equations of the discrete equation should reduce to the correct differential equations.

The key to success in the formulation and construction of best finite difference schemes is the application of the nonstandard modeling rules as given in Section 3.4. Also, it is equally important to be knowledgeable of both exact and best finite difference schemes for a large number of ordinary and partial differential equations.

For those few partial differential equations for which the general solution can be found, it is always the case that a functional relation exists between the space and time step-sizes, i.e.,

$$\Delta t = \Sigma(\Delta x). \qquad (7.5.1)$$

Unfortunately, most of the best schemes constructed in this chapter do not allow the determination of such a relation. Additional information is required to obtain such restrictions. The imposition of conditional stability often provides this functional relation for linear equations.

The partial differential equations investigated in this chapter have only quadratic nonlinearities. It would be of great value to generalize these procedures to other types of nonlinear terms and other classes of equations.

References

1. W. F. Ames, *Nonlinear Partial Differential Equations in Engineering* (Academic Press, New York, 1965).
2. G. B. Whitham, *Linear and Nonlinear Waves* (Wiley-Interscience, New York, 1974).
3. D. Potter, *Computational Physics* (Wiley-Interscience, New York, 1973).
4. F. B. Hildebrand, *Finite-Difference Equations and Simulations* (Prentice-Hall; Englewood Cliffs, NJ; 1968).
5. R. D. Richtmyer and K. W. Morton, *Difference Methods for Initial-Value Problems* (Interscience, New York, 2nd edition, 1967).
6. G. D. Smith, *Numerical Solution of Partial Differential Equations: Finite Difference Methods* (Clarendon Press, Oxford, 1978).
7. A. R. Mitchell and D. F. Griffiths, *Finite Difference Methods in Partial Differential Equations* (Wiley, New York, 1980).
8. R. E. Mickens, *Journal of Sound and Vibration* **100**, 452–455 (1985). Exact finite difference schemes for the nonlinear unidirectional wave equation.
9. R. E. Mickens, *Numerical Methods for Partial Differential Equations* **5**, 313–325 (1989). Exact solutions to a finite-difference model of a nonlinear reaction-advection equation: Implications for numerical analysis.

10. R. E. Mickens, *Journal of Sound and Vibration* **146**, 342–344 (1991). Analysis of a new finite-difference scheme for the linear advection-diffusion equation.

11. E. Zauderer, *Partial Differential Equations of Applied Mathematics* (Wiley-Interscience, New York, 1983).

12. R. E. Mickens, *Difference Equations: Theory and Applications* (Van Nostrand Reinhold, New York, 2nd edition, 1990).

13. R. Peyret and T. D. Taylor, *Computational Methods for Fluid Flow* (Springer-Verlag, New York, 1983).

14. T. F. Chan, *SIAM Journal of Numerical Analysis* **21**, 272–284 (1984). Stability analysis of finite-difference schemes for the advection-diffusion equation.

15. J. C. Strikwerda, *Finite Difference Schemes and Partial Differential Equations* (Wadsworth; Pacific Grove, CA; 1989).

16. L. R. Bentley, G. F. Pinder and I. Herrera, *Numerical Methods for Partial Differential Equations* **5**, 227–240 (1989). Solution of the advective-dispersive transport equation using a least squares collocation, Eulerian-Lagrangian method.

17. G. E. Forsythe and W. R. Wasow, *Finite-Difference Methods for Partial Differential Equations* (Wiley, New York, 1960).

18. R. E. Mickens, Pitfalls in the numerical integration of differential equations, in *Analytical Techniques for Material Characterization*, W. E. Collins et al., editors (World Scientific Publishing, Singapore, 1987), pp. 123–143.

19. R. E. Mickens, *Numerical Methods for Partial Differential Equations* **2**, 123–129 (1986). Exact solutions to difference equation models of Burgers' equation.

20. D. Greenspan and V. Casulli, *Numerical Analysis for Applied Mathematics, Science and Engineering* (Addison-Wesley; Redwood City, CA; 1988).

21. H. S. Carslaw and J. C. Jaeger, *Conduction of Heat in Solids* (Clarendon, London, 2nd edition, 1959).

22. J. Crank and P. Nicolson, *Proceedings of the Cambridge Philosophical Society* **43**, 50–67 (1947). A practical method for numerical evaluation of solutions of partial differential equations of heat conduction type.

23. M. C. Bhattacharya, *Communications in Applied Numerical Methods* **6**, 173–184 (1990). Finite-difference solutions of partial differential equations.

24. R. E. Mickens, *Numerical Methods for Partial Differential Equations* **7**, 299–302 (1991). Construction of a novel finite-difference scheme for a nonlinear diffusion equation.

25. R. A. Fisher, *Annuals of Eugenics* **7**, 355 (1937). The wave of advance of advantageous genes.

26. A. R. Mitchell and J. C. Bruch, Jr., *Numerical Methods for Partial Differential Equations* **1**, 13–23 (1985). A numerical study of chaos in a reaction-diffusion equation.

27. N. Parekh and S. Puri, *Physical Review E* **2**, 1415–1418 (1993). Velocity selection in coupled-map lattices.

28. J. B. Marion, *Classical Dynamics of Particles and Systems* (Academic Press, New York, 2nd edition, 1970).

29. R. E. Mickens, "New finite-difference scheme for the Fisher equation," Clark Atlanta University preprint; June 1993.

30. J. M. Burgers, *Advances in Applied Mechanics* **1**, 171–199 (1948). A mathematical model illustrating the theory of turbulence.

31. R. E. Mickens and J. N. Shoosmith, *Journal of Sound and Vibration* **142**, 536–539 (1990). A discrete model of a modified Burgers' partial differential equation.

32. R. E. Mickens, *Transactions of the Society for Computer Simulation* **8**, 109–117 (1991). Nonstandard finite difference schemes for partial differential equations.

Chapter 8

Schrödinger Differential Equations

8.1 Introduction

Schrödinger type ordinary and partial differential equations arise in the modeling of a large number of physical phenomena. Particular areas include quantum mechanics, ocean acoustics, optics, plasma physics and seismology [1–4]. A large literature exists on the determination of asymptotic techniques for calculating analytic approximations to the solutions of these equations [1, 5, 6]. The work presented in this chapter will center on the construction of finite difference techniques for use in the numerical integration of Schrödinger type ordinary and partial differential equations in one space dimension [7–9].

For our purposes, the Schrödinger partial differential equation takes the form

$$\frac{\partial u}{i\partial t} = \frac{\partial^2 u}{i\partial x^2} + f(x)u. \qquad (8.1.1)$$

This equation is usually called the time-dependent Schrödinger equation. The related Schrodinger time-independent ordinary differential equation can be expressed

$$\frac{d^2 u}{dx^2} + \bar{f}(x)u = 0, \qquad (8.1.2)$$

where, for many applications, $f(x)$ and $\bar{f}(x)$ differ only by a constant [1].

In Section 8.2, we discuss a novel finite difference scheme for Eq. (8.1.2) [9]. The generalization of this procedure to include the Numerov scheme [10] is then presented [11]. Section 8.3 begins with an examination of the difficulties of constructing stable finite difference schemes for the free-particle Schrödinger equation

$$\frac{\partial u}{i\partial t} = \frac{\partial^2 u}{\partial x^2}. \qquad (8.1.3)$$

We show that this problem can, in part, be overcome by using the concept of nonstandard discrete derivatives [7, 8]. Next, this result is used to construct a novel discrete model for the time-dependent Schrödinger equation. We end the section with an application of these ideas to the nonlinear, cubic Schrödinger equation.

8.2 Schrödinger Ordinary Differential Equations

8.2.1 *Numerov Method*

The simplest finite difference scheme for the time-independent Schrödinger equation

$$\frac{d^2u}{dx^2} + f(x)u = 0, \tag{8.2.1}$$

(where we have dropped the bar over the function $f(x)$, see Eq. (8.1.2)) is

$$\frac{u_{m+1} - 2u_m + u_{m-1}}{h^2} + f_m u_m = 0, \tag{8.2.2}$$

where

$$f_m = f(x_m), \quad x_m = (\Delta x)m = hm. \tag{8.2.3}$$

However, for a variety of reasons, including the possible existence of numerical instabilities and the lack of numerical accuracy, other discrete models have been considered [10, 12, 13]. A popular method is the Numerov algorithm [10]. The starting point for deriving this scheme is the relation [14]

$$\frac{u_{m+1} - 2u_m + u_{m-1}}{h^2} = u_m'' + \left(\frac{h^2}{12}\right)u_m'''' + O(h^4), \tag{8.2.4}$$

where

$$u_m = u(x_m), \quad f_m = f(x_m), \tag{8.2.5}$$

and

$$u_m'' = \frac{d^2u}{dx^2}\bigg|_{x=x_m}, \quad u_m'''' = \frac{d^4u}{dx^4}\bigg|_{x=x_m}. \tag{8.2.6}$$

From Eq. (8.2.1), we have

$$u_m'' = -f_m u_m. \tag{8.2.7}$$

Consequently,

$$\begin{aligned}
u_m'''' &= -\frac{d^2}{dx^2}[f(x)u]\big|_{x=x_m} \\
&= -\left[\frac{f_{m+1}u_{m+1} - 2f_m u_m + f_{m-1}u_{m-1}}{h^2}\right] + O(h^2).
\end{aligned} \tag{8.2.8}$$

Substitution of Eqs. (8.2.7) and (8.2.8) into Eq. (8.2.4) gives

$$\frac{u_{m+1} - 2u_m + u_{m-1}}{h^2} = -f_m u_m$$
$$- \left(\frac{h^2}{12}\right) \left[\frac{f_{m+1}u_{m+1} - 2f_m u_m + f_{m-1}u_{m-1}}{h^2}\right]$$
$$+ O(h^4). \tag{8.2.9}$$

Simplifying this expression and neglecting terms of $O(h^4)$ gives the Numerov algorithm

$$\left[1 + \frac{h^2 f_{m+1}}{12}\right] u_{m+1} - 2\left[1 + \frac{5h^2 f_m}{12}\right] u_m + \left[1 + \frac{h^2 f_{m-1}}{12}\right] u_{m-1} = 0. \tag{8.2.10}$$

8.2.2 *Mickens-Ramadhani Scheme*

The finite difference scheme proposed for the time-independent Schrödinger equation is based on the use of nonstandard modeling rules for the construction of discrete representations for differential equations; see Section 3.4 and the references [15, 16]. We begin with the exact difference scheme for

$$\frac{d^2 u}{dx^2} + \lambda u = 0 \tag{8.2.11}$$

where λ is a constant. It is given by the expression

$$\frac{u_{m+1} - 2u_m + u_{m-1}}{(4\lambda) \sin^2\left(\frac{h\sqrt{\lambda}}{2}\right)} + \lambda u_m = 0. \tag{8.2.12}$$

Note that this relation holds whether λ is positive or negative. This result follows directly from the use of the relation:

$$\sin(i\theta) = i \sinh(\theta). \tag{8.2.13}$$

The Mickens-Ramadhani scheme replaces the constant λ is Eq. (8.2.12) by the discrete form of the function $f(x)$, i.e.,

$$\lambda \to f_m = f(x_m). \tag{8.2.14}$$

This gives the scheme

$$\frac{u_{m+1} - 2u_m + u_{m-1}}{\left(\frac{4}{f_m}\right) \sin^2\left(\frac{h\sqrt{f_m}}{2}\right)} + f_m u_m = 0. \tag{8.2.15}$$

Using the trigonometric identity

$$2\sin^2\theta = 1 - \cos 2\theta, \tag{8.2.16}$$

we obtain

$$u_{m+1} + u_{m-1} = 2\left[\cos(h\sqrt{f_m})\right]u_m. \qquad (8.2.17)$$

For purposes of comparison, we rewrite the simple finite difference scheme of Eq. (8.2.2) in this form; it is

$$u_{m+1} + u_{m-1} = 2\left(1 - \frac{h^2 f_m}{2}\right)u_m. \qquad (8.2.18)$$

Asymptotic behavior of the solutions to difference equations can vary widely between two equations that seemingly have minor differences in their structure. Thus, it is of interest to compare the solutions of the Mickens-Ramadhani and standard schemes to that of the discrete version of a differential equation with known asymptotic solution. We select the zero-th order Bessel equation

$$\frac{d^2 w}{dx^2} + \left(\frac{1}{x}\right)\frac{dw}{dx} + w = 0. \qquad (8.2.19)$$

The transformation [5]

$$\sqrt{x}w = u, \qquad (8.2.20)$$

converts Eq. (8.2.19) to the equation

$$\frac{d^2 u}{dx^2} + \left(1 + \frac{1}{4x^2}\right)u = 0. \qquad (8.2.21)$$

Using the WKB procedure, the following asymptotic $(x \to \infty)$ solution is obtained [5]

$$u(x) = A\left[\sin(x) - \frac{\cos(x)}{8x}\right] + B\left[\cos(x) + \frac{\sin(x)}{8x}\right] + O\left(\frac{1}{x^2}\right), \qquad (8.2.22)$$

where A and B are arbitrary constants.

Discrete versions of the WKB method also exist for calculating the asymptotic solutions of linear second-order difference equations [6, 17, 18]. Applying these procedures to Eqs. (8.2.17) and (8.2.18) gives, respectively [9], for $f_m = [1 + (4x_m^2)^{-1}]$,

$$u_m^{(MR)} = A\left[\sin(x_m) - \frac{\cos(x_m)}{8x_m}\right]$$
$$+ B\left[\cos(x_m) + \frac{\sin(x_m)}{8x_m}\right] + O\left(\frac{1}{x_m^2}\right), \qquad (8.2.23)$$

$$u_m^{(S)} = A \left\{ \sin[\phi(h)x_m] - \frac{\beta(h)\cos[\phi(h)x_m]}{8x_m} \right\}$$

$$+ B \left\{ \cos[\phi(h)x_m] + \frac{\beta(h)\sin[\phi(h)x_m]}{8x_m} \right\}$$

$$+ O\left(\frac{1}{x_m^2}\right), \tag{8.2.24}$$

where

$$\phi(h) = \left(\frac{1}{h}\right) \tan^{-1}\left[\frac{\sqrt{4h^2 - h^4}}{2 - h^2}\right], \tag{8.2.25}$$

$$\beta(h) = h^2 \left[1 - \frac{h^2}{4}\right]^{1/2} + \frac{(1 - h^2/2)^2}{\sqrt{1 - h^2/4}}. \tag{8.2.26}$$

Note that the Mickens-Ramadhani scheme agrees with the exact result to terms of $O(x_m^{-2})$, while the standard scheme always disagrees with the exact answer for finite step-size h.

8.2.3 *Combined Numerov-Mickens Scheme*

A finite difference scheme for the time-independent Schrödinger equation that combines the Numerov and Mickens-Ramadhani schemes has been constructed and studied by Chen et al. [11]. This scheme has the form

$$\left[1 + \frac{h^2 f_{m+1}}{12}\right] u_{m+1} + \left[1 + \frac{h^2 f_{m-1}}{12}\right] u_{m+1}$$

$$= 2\left[\cos\left(h\sqrt{f_m}\right)\right]\left[1 + \frac{h^2 f_m}{12}\right] u_m. \tag{8.2.27}$$

They call the new discrete model the "combined Numerov-Mickens finite difference scheme" (CNMFDS). Examination of the Numerov, Mickens-Ramadhani and CNMFDS representations allows the following conclusions to be reached:

(i) The Mickens-Ramadhani scheme is (formally) of $O(h^2)$, while the Numerov scheme is $O(h^4)$.

(ii) The Mickens-Ramadhani scheme is an exact finite difference model for $f(x) = $ constant. This is not the situation for the Numerov scheme.

(iii) The CNMFDS is of $O(h^4)$, just like the Numerov scheme, and it is also an exact finite difference scheme for $f(x) = $ constant.

Numerical experiments were also done by Chen et al. [11]. Their general conclusion was that the CNMFDS has potential for use in practical calculations.

8.3   Schrödinger Partial Differential Equations

Finite difference schemes for Schrödinger partial differential equations generally separate into two classes: implicit and explicit formulations [19–22]. Most investigations have focused on implicit schemes because of the good stability properties that these schemes possess. (Stability is used here in the sense that small errors in the initial data do not grow as the discrete-time is increased [23, 24, 25].) However, a major difficulty with implicit schemes is the need to solve large sets of systems of complex-valued algebraic equations. In contrast, many explicit schemes, such as a simple forward-Euler scheme, are unconditionally unstable [19, 20]. However, explicit schemes are generally easier to implement and have fewer computer-storage requirements as compared to implicit schemes. In Section 8.3.1, we show that it is possible to construct explicit, forward-Euler schemes for the so-called free-particle Schrödinger equation. This construction is based on the use of a nonstandard denominator function for the discrete-time derivative. These schemes are conditionally stable. In Section 8.3.2, we apply these results and the use of nonstandard modeling rules to obtain finite difference nodels for the full time-dependent Schrödinger equation. Finally, in Section 8.3.3, an application is made of these procedures to the nonlinear, cubic Schrödinger equation.

8.3.1 $u_t = iu_{xx}$

The simplest Schrödinger type partial differential equation is the free-particle equation [1, 19, 20]

$$\frac{\partial u}{i\partial t} = \frac{\partial^2 u}{\partial x^2}. \qquad (8.3.1)$$

The direct forward-Euler scheme

$$\frac{u_m^{k+1} - u_m^k}{i\Delta t} = \frac{u_{m+1}^k - 2u_m^k + u_{m-1}^k}{(\Delta x)^2}, \qquad (8.3.2)$$

is unconditionally unstable for any choice of Δx and Δt [19, 20]. We now demonstrate that a conditionally stable finite difference model can be constructed using nonstandard modeling rules [7].

The following is a list of properties that an explicit finite difference model for Eq. (8.3.1) should possess:

(i) The discrete-time derivative must be of first-order.

(ii) The discrete-space derivative must be of second-order and centered about x_m.

(iii) There should not be any *ad hoc* terms in the scheme.

(iv) The finite difference scheme should be (at least) conditionally stable.

Requirements (i) and (ii) are consequences of Eq. (8.3.1) being a partial differential equation that is first-order in the time derivative and second-order in the space derivative. A higher order scheme for either discrete derivative would lead to numerical instabilities [15, 16]. Condition (iii) is the requirement that the finite difference scheme should be "natural," i.e., every term in the discrete model should have a counterpart in the differential equation. Finally, requirement (iv) is needed to ensure that practical calculations can actually be done using the scheme.

It should be indicated that these requirements automatically eliminate from consideration several discrete schemes for the Schrödinger partial differential equation. One example is the explicit finite difference scheme of Cahn et al. [19],

$$\frac{u_m^{k+1} - u_m^k}{i\Delta t} = \frac{u_{m+1}^k - 2u_m^k + u_{m-1}^k}{(\Delta x)^2}$$
$$+ (\alpha + i\beta)\Delta t \left[\frac{u_{m+2}^k - 4u_{m+1}^k + 6u_m^k - 4u_{m-1}^k + u_{m-2}^k}{(\Delta x)^2}\right], \quad (8.3.3)$$

where α and β are certain constants. This form is eliminated because of the *ad hoc* nature of the second term on the right-side of the equation. The central difference discrete-time form [26]

$$\frac{u_m^{k+1} - u_m^{k-1}}{2i\Delta t} = \frac{u_{m+1}^k - 2u_m^k + u_{m-1}^k}{(\Delta x)^2}, \quad (8.3.4)$$

is also eliminated since the partial difference equation is second-order in the discrete-time variable and thus violates condition (i).

Consider now the first-order ordinary differential equation

$$\frac{dw}{dt} = i\lambda w, \quad (8.3.5)$$

where λ is an arbitrary real number. The exact finite difference scheme for it is

$$\frac{w_{k+1} - w_k}{\left(\frac{e^{i\lambda h} - 1}{i\lambda}\right)} = i\lambda w_k, \quad h = \Delta t. \quad (8.3.6)$$

A conventional discrete model for Eq. (8.3.5) is

$$\frac{w_{k+1} - w_k}{h} = i\lambda w_k. \quad (8.3.7)$$

Note that the denominator function, $D(h, \lambda)$, for the exact scheme is

$$D(h, \lambda) = \frac{e^{i\lambda h} - 1}{i\lambda} = h + i\left(\frac{\lambda h^2}{2}\right) + O(\lambda^2 h^2), \qquad (8.3.8)$$

in contrast to the simple denominator function, h, for Eq. (8.3.7). This result suggests the following form for an explicit scheme for Eq. (8.3.1)

$$\frac{u_m^{k+1} - u_m^k}{iD_1(\Delta t, \lambda)} = \frac{u_{m+1}^k - 2u_m^k + u_{m-1}^k}{D_2(\Delta x, \lambda)}, \qquad (8.3.9)$$

where the denominator functions have the properties

$$D_1(\Delta t, \lambda) = \Delta t + i\lambda(\Delta t)^2 + O[(\Delta t)^3], \qquad (8.3.10)$$

$$D_2(\Delta x, \lambda) = (\Delta x)^2 + O[(\Delta x)^4]. \qquad (8.3.11)$$

At this stage of the analysis, the exact dependencies of D_1 and D_2 on $(\Delta x, \Delta t, \lambda)$ do not have to be specified in any more detail than that given by the conditions of Eqs. (8.3.10) and (8.3.11). These requirements ensure that the finite difference scheme is both convergent to and consistent with the original partial differential equation.

Define $R(\Delta t, \Delta x, \lambda)$ to be

$$R \equiv \frac{iD_1(\Delta t, \lambda)}{D_2(\Delta x, \lambda)}. \qquad (8.3.12)$$

With this definition Eq. (8.3.9) can be rewritten to the form

$$u_m^{k+1} = Ru_{m+1}^k + (1 - 2R)u_m^k + Ru_{m-1}^k. \qquad (8.3.13)$$

We now need to do a stability analysis for this finite difference scheme. The stability concept to be applied is the von Neumann or Fourier-series method [23–25]. This procedure is based on the fact that both Eqs. (8.3.1) and (8.3.9) are linear equations and the physical solutions of Eq. (8.3.1) are bounded for all times. With this in mind, the stability properties of Eq. (8.3.13) can be studied by considering a typical Fourier mode

$$u_m^k = C(k)e^{i\omega(\Delta x)x}, \qquad (8.3.14)$$

where ω is a constant, and requiring that $C(k)$ be bounded for all k. This concept of stability is called practical stability [19, 23, 24].

The substitution of Eq. (8.3.14) into Eq. (8.3.13) gives the following equation for $C(k)$:

$$C(k + l) = AC(k) \qquad (8.3.15)$$

where

$$A = 1 - 4R \sin^2 \left(\frac{\theta}{2} \right), \quad \theta = \omega(\Delta x). \tag{8.3.16}$$

The solution to this first-order difference equation is

$$C(k) = C(0) A^k, \tag{8.3.17}$$

where $C(0)$ is an arbitrary constant. Now, if $C(k)$ is to be bounded, then A must satisfy the condition

$$|A| \leq 1. \tag{8.3.18}$$

Let R, see Eq. (8.3.12), be written as

$$R = R_1 + iR_2, \tag{8.3.19}$$

where R_1 and R_2 are real functions of Δt, Δx and λ. A straightforward calculation shows that the inequality of Eq. (8.3.18) is equivalent to the expression

$$R_1^2 + R_2^2 \leq \frac{R_1}{2}, \tag{8.3.20}$$

or

$$\left(R_1 - \frac{1}{4} \right)^2 + R_2^2 \leq \frac{1}{16}. \tag{8.3.21}$$

This inequality has the following geometric interpretation: In the (R_1, R_2) plane, the finite difference scheme of Eq. (8.3.13) is stable for all points on and inside the semi-circle of radius 0.25, centered at $(0.25, 0)$, and lying in the upper plane. We will refer to this relation as the *circle condition* [7, 27].

In practice, the circle condition is to be used as follows:

(a) Select denominator functions with the properties given by Eqs. (8.3.10) and (8.3.11).

(b) Calculate $R(\Delta t, \Delta x, \lambda)$ as given by Eq. (8.3.12).

(c) Calculate R_1 and R_2, respectively, the real and imaginary parts of R.

(d) Select a point $(\bar{R}_1, \bar{R}_2)$ consistent with the circle condition of Eq. (8.3.21).

(e) Set $R_1(\Delta t, \Delta x, \lambda)$ and $R_2(\Delta t, \Delta x, \lambda)$ equal, respectively, to $\bar{R}_1$ and $\bar{R}_2$, i.e.,

$$R_1(\Delta t, \Delta t, \lambda) = \bar{R}_1, \quad R_2(\Delta t, \Delta t, \lambda) = \bar{R}_2. \tag{8.3.22}$$

(f) Choose a value for the space step-size, Δx, and solve Eqs. (8.3.22) for Δt and λ in terms of Δx. Carrying out these operations gives

$$\Delta t = f_1(\Delta x), \quad \lambda = f_2(\Delta x). \tag{8.3.23}$$

Therefore, the selection of the point $(\bar{R}, \bar{R}_2)$, satisfying the circle condition and the relations of Eq. (8.3.23), completely defines the explicit finite difference scheme given by Eq. (8.3.9) or (8.3.13).

The Schrödinger equation, in general, has solutions for which the "amplitude" does not decrease with the increase of time. In fact, for physical applications, the solutions to Eq. (8.3.1) satisfy the requirement [1]

$$\int_{-\infty}^{\infty} |u(x,t)|^2 dx = 1. \tag{8.3.24}$$

Therefore, the circle condition of Eq. (8.3.21) should read

$$\left(R_1 - \frac{1}{4}\right)^2 + R_2^2 = \frac{1}{16}. \tag{8.3.25}$$

Let us apply this result to the simple finite difference scheme of Eq. (8.3.2). For this case, we have

$$D_1 = \Delta t, \quad D_2 = (\Delta x)^2, \tag{8.3.26}$$

$$R = \frac{i\Delta t}{(\Delta x)^2}, \tag{8.3.27}$$

$$R_1 = 0, \quad R_2 = \frac{\Delta t}{(\Delta x)^2}. \tag{8.3.28}$$

Substitution of these values into the circle condition of Eq. (8.3.21) gives

$$R_2^2 = 0. \tag{8.3.29}$$

This cannot be satisfied for any finite values of Δt and Δx. Consequently, we conclude that the simple forward-Euler scheme is unstable. This is the same conclusion that is reached by the usual methods of stability analysis [19, 20].

Let us now consider a finite difference scheme for Eq. (8.3.1) such that [7, 8]

$$D_1 = \Delta t + i\lambda(\Delta t)^2, \quad D_2 = (\Delta x)^2. \tag{8.3.30}$$

This corresponds to the use of a nonstandard discrete time-derivative and a standard discrete space-derivative. Therefore,

$$R = -\lambda \left(\frac{\Delta t}{\Delta x}\right)^2 + i\left[\frac{\Delta t}{(\Delta x)^2}\right], \tag{8.3.31}$$

with

$$R_1 = -\lambda \left(\frac{\Delta t}{\Delta x}\right)^2, \quad R_2 = \frac{\Delta t}{(\Delta x)^2}. \tag{8.3.32}$$

Selection of the (circle condition) point $(\bar{R}_1, \bar{R}_2)$

$$\bar{R}_1 = \frac{1}{4}, \quad \bar{R}_2 = \frac{1}{4}, \tag{8.3.33}$$

gives

$$-\lambda \left(\frac{\Delta t}{\Delta x}\right)^2 = \frac{1}{4}, \quad \frac{\Delta t}{(\Delta x)^2} = \frac{1}{4}. \tag{8.3.34}$$

These equations can be solved for Δt and λ in terms of Δx. Doing this gives

$$\Delta t = \frac{(\Delta x)^2}{4}, \quad \lambda = -\frac{4}{(\Delta x)^2}, \tag{8.3.35}$$

and

$$D_1 = \left(\frac{1-i}{4}\right)(\Delta x)^2, \quad R = \frac{1+i}{4}. \tag{8.3.36}$$

Consequently, our nonstandard explicit finite difference scheme for the free-particle Schrödinger partial differential equation is [7]

$$\frac{u_m^{k+1} - u_m^k}{i\left(\frac{1-i}{4}\right)(\Delta x)^2} = \frac{u_{m+1}^k - 2u_m^k + u_{m-1}^k}{(\Delta x)^2}, \tag{8.3.37}$$

or

$$\frac{u_m^{k+1} - u_m^k}{i\left(\frac{1-i}{4}\right)} = u_{m+1}^k - 2u_m^k + u_{m-1}^k, \tag{8.3.38}$$

with

$$\Delta t = \frac{(\Delta x)^2}{4}. \tag{8.3.39}$$

The above scheme is called conditionally stable since practical stability [19, 23] holds only if the functional relation of Eq. (8.3.39) is satisfied for the step-sizes.

For completeness, we now investigate the stability properties of two other standard finite difference models for Eq. (8.3.1). The use of a central-difference discrete time-derivative gives the scheme

$$\frac{u_m^{k+1} - u^{k-1}}{2i\Delta t} = \frac{u_{m+1}^k - 2u_m^k + u_{m-1}^k}{(\Delta x)^2}. \tag{8.3.40}$$

Let β be defined as

$$\beta = \frac{2i\Delta t}{(\Delta x)^2} = i\bar{\beta}. \tag{8.3.41}$$

Using this, Eq. (8.3.40) can be written as

$$u_m^{k+1} = u_m^{k-1} + i\bar{\beta}(u_{m+1}^k - 2u_m^k + u_{m-1}^k). \tag{8.3.42}$$

Substituting

$$u_m^k = C(k)e^{i\omega(\Delta x)m} = C(k)e^{i\theta m} \tag{8.3.43}$$

into Eq. (8.3.42) gives the following linear, second-order difference equation for $C(k)$:

$$C(k+1) + [2i\bar{\beta}(1 - \cos\theta)]C(k) - C(k-1) = 0. \tag{8.3.44}$$

The solution to this equation has the form [28]

$$C(k) = B_1(r_1)^k + B_2(r_2)^k, \tag{8.3.45}$$

where r_1 and r_2 are solutions to the characteristic equation

$$r^2 + [2i\bar{\beta}(1 - \cos\theta)]r - 1 = 0. \tag{8.3.46}$$

Therefore,

$$r_{1,2} = -i\bar{\beta}(1 - \cos\theta) \pm \sqrt{1 - \bar{\beta}^2(1 - \cos\theta)^2}, \tag{8.3.47}$$

and

$$|r_1| = 1, \quad |r_2| = 1. \tag{8.3.48}$$

These results imply that $C(k)$ oscillates with a constant amplitude. Hence, the scheme given by Eq. (8.3.40) is unconditionally stable. In other words, this scheme has practical stability for all values of Δx and Δt. Note, however, that this discrete model is eliminated by our nonstandard modeling rules since the discrete time-derivative is of second-order.

Now consider a standard implicit finite difference scheme. This is given by the expression

$$\frac{u_m^{k+1} - u_m^k}{i\Delta t} = \frac{u_{m+1}^{k+1} - 2u_m^{k+1} + u_{m-1}^{k+1}}{(\Delta x)^2}. \tag{8.3.49}$$

Defining β_1 to be

$$\beta_1 = \frac{i\Delta t}{(\Delta x)^2}, \tag{8.3.50}$$

we have

$$u_m^{k+1} = \beta_1(u_{m+1}^{k+1} + u_{m-1}^{k+1}) - 2\beta_1 u_m^{k+1} + u_m^k. \qquad (8.3.51)$$

Substituting u_m^k from Eq. (8.3.43) into this equation gives the following linear first-order difference equation for $C(k)$:

$$[1 + 2\beta_1(1 - \cos\theta)]C(k + 1) = C(k). \qquad (8.3.52)$$

Since

$$|[1 + 2\beta_1(1 - \cos\theta)]| \geq 1, \qquad (8.3.53)$$

it can be concluded that the finite difference scheme of Eq. (8.3.49) is unconditionally stable for all values of Δx and Δt. However, in practice, this scheme should not be used since the amplitude, $C(k)$, generally decreases with the increase of the discrete time variable.

We conclude this section with a demonstration that an exact, explicit finite difference scheme does not exist for the free-partial Schrödinger equation.

First, we require that the discrete model satisfy the two restrictions:
(i) The discrete time-derivative be of first-order.
(ii) The discrete space-derivative be a centered, second-order expression.
Since Eq. (8.3.1) is a linear partial differential equation having only two terms, it can always be solved by the method of separation of variables. The corresponding finite difference scheme must also have this property. Therefore, any proposed exact finite difference scheme can be "checked" by also calculating its solutions by the method of separation of variables [27]. If the discrete model is an exact scheme, then we will find that

$$u_m^k = u(x_m, t_k) \qquad (8.3.54)$$

where u_m^k is the solution of the finite difference equation and $u(x, t)$ is the solution to the differential equation.

The substitution of

$$u(x, t) = X(x)T(t) \qquad (8.3.55)$$

into Eq. (8.3.1) gives

$$\frac{X''}{X} = \frac{T'}{iT} = -i\omega^2, \qquad (8.3.56)$$

where the primes denote the order of the indicated derivatives and ω^2 is the separation constant. For given ω^2, the special solution $u_\omega(x, t)$ is

$$u_\omega(x, t) = A(\omega)e^{i\omega(x - \omega t)}, \qquad (8.3.57)$$

and the general solution is given by "summation," i.e.,

$$u(x,t) = \sum_\omega u_\omega(x,t)d\omega. \tag{8.3.58}$$

The general form of an explicit scheme, for Eq. (8.3.1), that satisfies the above two requirements is

$$\frac{u^{k+1} - \Gamma(h,\ell^2)u_m^k}{ih\Phi(h,\ell^2)} = \frac{(u_{m+1}^k + u_{m-1}^k)P(h,\ell^2) - 2Q(h,\ell^2)u_m^k}{\ell^2\Sigma(h,\ell^2)}, \tag{8.3.59}$$

where $h = \Delta t$, $\ell = \Delta x$, and the functions appearing in Eq. (8.3.59) have the representations:

$$\Phi(h,\ell^2) = 1 + h\Phi_1(h,\ell^2), \tag{8.3.60}$$

$$\Sigma(h,\ell^2) = 1 + \ell^2\Sigma_1(h,\ell^2), \tag{8.3.61}$$

$$\Gamma(h,\ell^2) = 1 + h\Gamma_1(h,\ell^2), \tag{8.3.62}$$

$$P(h,\ell^2) = 1 + \ell^2 P_1(h,\ell^2), \tag{8.3.63}$$

$$Q(h,\ell^2) = 1 + \ell^2 Q_1(h,\ell^2). \tag{8.3.64}$$

Note that $\Phi(h,\ell^2)$ may be a complex-valued function. Equation (8.3.59) can be rewritten as

$$u_m^{k+1} = A(u_{m+1}^k + u_{m-1}^k) + Bu_m^k, \tag{8.3.65}$$

where

$$A = \frac{ih\Phi P}{\ell^2\Sigma}, \tag{8.3.66}$$

$$B = \Gamma - \frac{2i\Phi Q}{\ell^2\Sigma}. \tag{8.3.67}$$

Now assume a separation-of-variables solution

$$u_m^k = C^k D_m, \tag{8.3.68}$$

and substitute this into Eq. (8.3.65) to obtain

$$\frac{C^{k+1}}{C^k} = \frac{A(D_{m+1} + D_{m-1}) + BD_m}{D_m} = \alpha, \tag{8.3.69}$$

where α is the separation constant. Now require that C^k has the form

$$C^k = e^{i\omega^2(\Delta t)k}. \tag{8.3.70}$$

This means that the separation constant α is

$$\alpha = e^{i\omega^2 h}, \quad h = \Delta t. \tag{8.3.71}$$

Therefore, D_m satisfies the following second-order linear difference equation

$$D_{m+1} + 2 \left[\frac{B - e^{i\omega^2 h}}{2A} \right] D_m + D_{m-1} = 0. \tag{8.3.72}$$

For an exact scheme, this equation must have the solution

$$D_m = e^{i\omega(\Delta x)m}. \tag{8.3.73}$$

However, this requires that

$$\frac{B - e^{i\omega^2 h}}{2A} = -\cos(\omega\ell), \quad \ell = \Delta x. \tag{8.3.74}$$

Written out in detail, the last equation is

$$\frac{\ell^2 \Sigma e^{i\omega^2 h} + 2ih\Phi Q - \ell\Gamma\Sigma}{2ih\Phi P} = \cos(\omega\ell). \tag{8.3.75}$$

In general, for arbitrary values of ω^2, the two sides of this expression are not equal. Consequently, we conclude that the free-particle Schrödinger equation does not have an exact, explicit finite difference scheme [28].

8.3.2 $u_t = i[u_{xx} + f(x)u]$

The time-dependent Schrödinger equation is

$$\frac{\partial u}{i\partial t} = \frac{\partial^2 u}{\partial x^2} + f(u)u. \tag{8.3.76}$$

Based on the discussion of the previous section, we expect a discrete model for Eq. (8.3.76) to take the form [8, 29]

$$\frac{u_m^{k+1} - u_m^k}{D_1(\lambda, \Delta t, \Delta x, f_m)} = \frac{u_{m+1}^k - 2u_m^k + u_{m-1}^k}{D_2(\lambda, \Delta t, \Delta x, f_m)} + f_m u_m^k, \tag{8.3.77}$$

where

$$f_m = f(x_m), \quad x_m = (\Delta x)m, \tag{8.3.78}$$

and λ is to be selected such that the scheme is conditionally stable. The following denominator functions are suitable for this purpose:

$$D_1 = e^{i\lambda(\Delta t)} \left[\frac{e^{i(\Delta t)f_m} - 1}{f_m} \right], \tag{8.3.79}$$

$$D_2 = \left(\frac{4}{f_m} \right) \sin^2 \left[\frac{(\Delta x)\sqrt{f_m}}{2} \right]. \tag{8.3.80}$$

Examination of this scheme shows that it has the following properties:

(i) The discrete time-independent part of the scheme,

$$\frac{u_{m+1} - 2u_m + u_{m-1}}{\left(\frac{4}{f_m}\right) \sin^2\left[\frac{(\Delta x)\sqrt{f_m}}{2}\right]} + f_m u_m = 0, \tag{8.3.81}$$

reduces to the best difference scheme for the ordinary differential equation

$$\frac{d^2 u}{dx^2} + f(x)u = 0. \tag{8.3.82}$$

(See Section 8.2.2.)

(ii) Taking the limit

$$f_m \to 0, \tag{8.3.83}$$

gives

$$\frac{u_m^{k+1} - u_m^k}{i(\Delta t)e^{i\lambda(\Delta t)}} = \frac{u_{m+1}^k - 2u_m^k + u_{m-1}^k}{(\Delta x)^2}, \tag{8.3.84}$$

which is a best difference scheme for

$$\frac{\partial u}{i\partial t} = \frac{\partial^2 u}{\partial x^2}. \tag{8.3.85}$$

The circle condition of Section 8.3.1 allows us to determine that this scheme is conditionally stable if, for example,

$$\Delta t = \frac{(\Delta x)^2}{2\sqrt{2}}, \quad \lambda = \frac{7\pi}{\sqrt{2}(\Delta x)^2}. \tag{8.3.86}$$

Note that

$$\lambda(\Delta t) = \frac{7\pi}{4}, \tag{8.3.87}$$

and

$$e^{i\lambda(\Delta t)} = e^{-i\pi/4} = \frac{1-i}{\sqrt{2}}. \tag{8.3.88}$$

We assume that these conditions hold for the full finite difference scheme given by Eqs. (8.3.77), (8.3.79) and (8.3.80).

Other functional forms can be chosen for the discrete-time denominator function D_1. However, the $f_m \to 0$ limit will always give relations similar to those stated in Eqs. (8.3.86) and (8.3.87).

8.3.3 Nonlinear, Cubic Schrödinger Equation

There exists a vast literature on both the properties and the numerical integration of the nonlinear, cubic Schrödinger partial differential equation [30–35]. This equation has the form

$$\frac{\partial u}{i \partial t} = \frac{\partial^2 u}{\partial x^2} + |u|^2 u, \tag{8.3.89}$$

and describes the asymptotic limiting behavior of a slowly varying dispersive wave envelope traveling in a nonlinear medium. Our purpose in this section is to construct a nonstandard, explicit finite difference scheme for Eq. (8.3.89).

We begin by again considering the Duffing ordinary differential equation

$$\frac{d^2 u}{dx^2} + \lambda u + u^3 = 0. \tag{8.3.90}$$

A best finite difference scheme for it is (see Section 5.3)

$$\frac{u_{m+1} - 2u_m + u_{m-1}}{\left(\frac{4}{\lambda}\right) \sin^2 \left[\frac{(\Delta x)\sqrt{\lambda}}{2}\right]} + \lambda u_m + u_m^2 \left(\frac{u_{m+1} + u_{m-1}}{2}\right) = 0. \tag{8.3.91}$$

Taking the limit $\lambda \to 0$, we obtain

$$\frac{u_{m+1} - 2u_m + u_{m-1}}{(\Delta x)^2} + u_m^2 \left(\frac{u_{m+1} + u_{m-1}}{2}\right) = 0, \tag{8.3.92}$$

as a best scheme for the equation

$$\frac{d^2 u}{dx^2} + u^3 = 0. \tag{8.3.93}$$

These results suggest that for the differential equation

$$\frac{d^2 u}{dx^2} + |u|^2 u = 0, \tag{8.3.94}$$

where u is now a complex-valued function, a possible best scheme is the expression

$$\frac{u_{m+1} - 2u_m + u_{m-1}}{(\Delta x)^2} + u_m u_m^* \left(\frac{u_{m+1} + u_{m-1}}{2}\right) = 0. \tag{8.3.95}$$

Combining all these results, we obtain the following explicit finite difference scheme for the nonlinear, cubic Schrödinger equation

$$\frac{u_m^{k+1} - u_m^k}{i D_1(\Delta t), \Delta x, \lambda)} = \frac{u_{m+1}^k - 2u_m^k + u_{m-1}^k}{(\Delta x)^2} + (u_m^k)^* \left(\frac{u_{m+1}^k + u_{m-1}^k}{2}\right) u_m^{k+1}, \tag{8.3.96}$$

where D_1 is a suitable denominator function. If D_1 is selected, such that the linear part of Eq. (8.3.96) satisfies the circle condition, then from Eqs. (8.3.35) and (8.3.36)

$$iD_1 = \left(\frac{1+i}{4}\right)(\Delta x)^2, \qquad (8.3.97)$$

$$\Delta t = \frac{(\Delta x)^2}{4}, \qquad (8.3.98)$$

and Eq. (8.3.96) becomes

$$u_m^{k+1} = \frac{\delta(u_{m+1}^k + u_{m-1}^k) + (1 - 2\beta)u_m^k}{1 - \beta(\Delta x)^2(u_m^k)^* \left[\frac{u_{m+1}^k + u_{m-1}^k}{2}\right]}, \qquad (8.3.99)$$

$$\beta = \frac{1+i}{4}. \qquad (8.3.100)$$

In summary, the discrete model of the nonlinear, cubic Schrödinger equation given by Eq. (8.3.96) or Eqs. (8.3.99) and (8.3.100), is constructed by applying the nonstandard modeling rules as presented and discussed in Sections 3.4 and 3.5. These rules lead to an essentially unique structure for the finite difference scheme.

References

1. E. Merzbacher, *Quantum Mechanics* (Wiley, New York, 1961).
2. F. Herman and E. Skillman, *Atomic Structure Calculations* (Prentice-Hall; Englewood Cliffs, NJ; 1963).
3. L. Brekhovskikh and Yu. Lupanov, *Fundamentals of Ocean Acoustics* (Springer-Verlag, New York, 1982).
4. A. K. Ghatak and K. Thyagarajan, *Contemporary Optics* (Plenum, New York, 1978).
5. P. B. Kahn, *Mathematical Methods for Scientists and Engineers: Linear and Nonlinear Systems* (Wiley-Interscience, New York, 1990).
6. C. M. Bender and S. A. Orszag, *Advanced Mathematical Methods for Scientists and Engineering* (McGraw-Hill, New York, 1978).
7. R. E. Mickens, *Physical Review A* **39**, 5508–5511 (1989). Stable explicit schemes for equations of Schrödinger type.

8. R. E. Mickens, Construction of stable explicit finite-difference schemes for Schrödinger type differential equations, in *Computational Acoustics* - Volume I, D. Lee, A. Cakmak and R. Vichnevetsky, editors (Elsevier, Amsterdam, 1990), pp. 11–16.

9. R. E. Mickens and I. Ramadhani, *Physical Review A* **45**, 2074–2075 (1992). Finite-difference scheme for the numerical solution of the Schrödinger equation.

10. S. E. Koonin, *Computational Physics* (Addison-Wesley; Redwood City, CA; 1986). See Section 3.1.

11. R. Chen, Z. Xu and L. Sun, *Physical Review E* **47**, 3799–3802 (1993). Finite difference scheme to solve Schrödinger equations.

12. L. Gr. Ixaru and M. Rizea, *Journal of Computational Physics* **73**, 306–324 (1987). Numerov method maximally adapted to the Schrödinger equation.

13. A. D. Raptis and J. R. Cash, *Computer Physics Communications* **44**, 95–103 (1987). Exponential and Bessel fitting methods for the numerical solution of the Schrödinger equation.

14. F. B. Hildebrand, *Finite-Difference Equations and Simulations* (Prentice-Hall; Englewood Cliffs, NJ; 1968).

15. R. E. Mickens, Mathematical modeling of differential equations by difference equations, in *Proceedings of the First IMAC Conference on Computational Acoustics*, D. Lee, R. L. Steingberg and M. H. Schultz, editors (North-Holland, Amsterdam, 1987), pp. 387–393.

16. R. E. Mickens, *Numerical Methods for Partial Differential Equations* **2**, 313–325 (1989). Exact solutions to a finite-difference model for a nonlinear reaction-advection equation: Implications for numerical analysis.

17. R. B. Dingle and G. J. Morgan, *Applied Scientific Research* **18**, 221 (1967). WKB methods for differential equations.

18. R. E. Mickens and I. Ramadhani, WKB procedures for Schrödinger type difference equations, to appear in *Proceedings of the First World Congress of Nonlinear Analysts* (Tampa, FL; August 19–26, 1992).

19. T. F. Chan, D. Lee and L. Shen, *SIAM Journal of Numerical Analysis* **23**, 274–281 (1986). Stable explicit schemes for equations of the Schrödinger type.

20. D. Lee and S. T. McDaniel, *Ocean Acoustic Propagation by Finite Difference Methods* (Pergamon, Oxford, 1988).

21. T. F. Chan and L. Shen, *SIAM Journal of Numerical Analysis* **24**, 336–349 (1987). Stability analysis of difference schemes for variable coefficient Schrödinger type equations.

22. P. K. Chattaraj, S. R. Koneru and B. M. Deb, *Journal of Computational Physics* **72**, 504–512 (1987). Stability analysis of finite difference schemes for quantum mechanical equations of motion.

23. R. D. Richtmyer and K. W. Morton, *Difference Methods for Initial-Value Problems* (Interscience, New York, 1967).

24. L. Lapidus and G. F. Pinder, *Numerical Solution of Partial Differential Equations in Science and Engineering* (Wiley, New York, 1982).

25. G. D. Smith, *Numerical Solution of Partial Differential Equations: Finite Difference Methods* (Clarendon, Oxford, 2nd edition, 1978).

26. A. Goldberg, H. M. Schey and J. L. Schwartz, *American Journal of Physics* **35**, 177–186 (1967). Computer-generated motion pictures of one-dimensional quantum mechanical transmission and reflection phenomena.

27. R. E. Mickens, *Difference Equations: Theory and Applications* (Van Nostrand Reinhold, New York, 2nd edition, 1990).

28. R. E. Mickens, Proof of the impossibility of constructing exact finite-difference schemes for Schrödinger-type PDE's. Paper presented at the International Conference on Theoretical and Computational Acoustics (Mystic, Connecticut; July 5–9, 1993).

29. R. E. Mickens, *Computer Physics Communications* **63**, 203–208 (1991). Novel explicit finite-difference schemes for time-dependent Schrödinger equations.

30. J. M. Sanz-Serna and I. Christie, *Journal of Computational Physics* **67**, 348–360 (1986). A simple adaptive technique for nonlinear wave problems.

31. K. A. Ross and C. J. Thompson, *Physica* **135A**, 551–558 (1986). Iteration of some discretizations of the nonlinear Schrödinger equation.

32. B. M. Herbst and M. J. Ablowitz, *Physical Review Letters* **62**, 2065–2068 (1989). Numerically induced chaos in the nonlinear Schrödinger equation.

33. M. J. Ablowitz and H. Segur, *Solitons and the Inverse Scattering Transform* (Society for Industrial and Applied Mathematics, Philadelphia, 1981).

34. G. B. Whitham, *Linear and Nonlinear Waves* (Wiley-Interscience, New York, 1974).

35. R. K. Dodd, J. C. Eilbeck, J. D. Gibbon and H. C. Morris, *Solitons and Nonlinear Wave Equations* (Academic Press, New York, 1982).

Chapter 9

The NSFD Methodology

9.1 Introduction

The main purpose of the Chapter is to discuss (not explain, in the sense of a fundamental mathematical theory) the NSFD methodology and the steps required to construct valid discretizations based on this methodology. We also hope to correct some of the misconceptions related to just what is a NSFD scheme.

In a fundamental sense, the genesis of the NSFD methodology starts with the realization that the standard numerical procedures for differential equations may not be the proper or optimal procedure for the determination of finite difference models for these equations. It must be indicated that while the constraints imposed by the NSFD methodology may appear to make the construction of the discretizations more difficult to construct, the mere existence of such a formalism forces the conclusion that the standard numerical procedures must be extended and generalized. However, it should be clearly noted that these new types of schemes must not only produce "meaningful" numerical solutions, they must also insure the absence of nonsensical results such as oscillations of the solutions, the addition of extra fixed-points, changes in the stability of solutions, etc.

An additional point to understand is that the general NSFD methodology cannot be derived from the mathematical framework used to analyze the standard numerical procedures. The NSFD methodology has its own fundamental postulates and these are consistent with each other. Essentially, there are only three concepts which form the backbone of these procedures:

(i) dynamic consistency ... ,
(ii) construction of denominator functions ... ,

(iii) nonlocal representations of functions.

How the issues related to these three concepts work out for a particular system of differential equations depends on the overall mathematical structure of the equations. As will be seen, *the separate term by term discretization will not, in general, produce a valid NSFD scheme.*

This Chapter is organized as follows: Section 9.2 gives a brief summary of the modeling process. Section 9.3 provides a discussion of intrinsic time and space scales and relates these quantities to restrictions on the corresponding step-sizes. The concept of dynamic consistency is explained in Section 9.4, and its relevance is tied to the existence of numerical instabilities in Section 9.5. The calculation of denominator functions is presented in Section 9.6. Likewise, in Section 9.7, we illustrate the nonlocal discretization of functions. An important technique for constructing NSFD schemes is the method of sub-equations and in Section 9.8, we define this concept. Section 9.9 illustrates the explicit construction of the NSFD schemes for several elementary, but important differential equations. Finally, in the last section, we summarize what has been accomplished in the Chapter and make several comments related to the NSFD methodology.

9.2   The Modeling Process

A model of a "physical system" is the construction of an appropriate approximate mathematical representation of the system. Note the important fact that the considered physical system is but a sub-system of the universe and is taken to be effectively isolated from the remainder of the universe. Various interactions with the sub-system allows data to be collected and then analyzed to obtain patterns related to its important features.

It is from the patterns derived from examining the data that a mathematical model can be constructed. (After all, mathematics is the study of all conceivable patterns.) Next, the mathematical model should be thoroughly investigated to determine both the qualitative and quantitative properties of the differential equations and the expected important features of their solutions. Observe that this can be done without the benefit of knowing the exact solutions. In fact, there should be little expectation of ever having available exact analytical expressions for the general solutions. This is a major reason for the creation of various numerical techniques. A further complication is that the mathematical equations are not unique. There may be more than one set of equations, each of which gives an adequate, at this stage, model of the phenomena as verified by future analysis of the mathematical model.

Once a given mathematical model has been selected, the next step is to construct a finite difference discretization, followed by calculating the associated numerical solutions.

The numerical solutions and their representations as figures, graphs, empirical formulas, etc., are then compared to the data. Depending on the "quality" of this comparison and other requirements, the process stated above may have to be iterated a number of times to achieve the desired outcomes related to understanding and controlling the sub-system of interest.

9.3 Intrinsic Time and Space Scales

Every physical system has time, space, and dependent variables scales; see Mickens [1, Section 7.7.1]. To illustrate what this means, we consider three explicit examples.

Example A: Logistic equation

The logistic differential equation provides an important elementary model for the growth of a population, $u(t)$; where $u(t)$ is the population at time t. This equation is

$$\frac{du}{dt} = \lambda_1 u - \lambda_2 u^2, \quad u(0) = u_0 > 0. \tag{9.3.1}$$

Denote the physical units of $(u, t, \lambda_1, \lambda_2)$ by

$$[t] = T, \quad [u] = \#, \quad [\lambda_1] = \frac{1}{T}, \quad [\lambda_2] = \frac{1}{\#T}, \tag{9.3.2}$$

where T is a physical unit of time (second, minute, etc.), and $\#$ is the "unit" of measuring the population.

Now construct dimensional variables

$$u = U\bar{v}, \quad t = T\bar{t}, \tag{9.3.3}$$

where $(\bar{v}, \bar{t})$ are dimensionless variables. We are to determine the (constant) scales, U and T, and this can be done by replacing (u, t) in Eq. (9.3.1) by the expressions in Eq. (9.3.2); doing this gives

$$\left(\frac{U}{T}\right)\frac{d\bar{v}}{d\bar{t}} = (\lambda_1 U)\bar{v} - (\lambda_2 U^2)\bar{v}^2. \tag{9.3.4}$$

Dividing each term by (U, T) gives

$$\frac{d\bar{v}}{d\bar{t}} = (\lambda_1 T)\bar{v} - (\lambda_2 UT)\bar{v}^2. \tag{9.3.5}$$

If the coefficients on the right are set to one, then (T, U) are determined to be

$$T = \frac{1}{\lambda_1}, \quad U = \frac{\lambda_1}{\lambda_2}. \tag{9.3.6}$$

Note that U is the stable equilibrium state achieved by the population as time becomes unbounded. This time T is the intrinsic time scale of the population and its value is determined by λ_1.

We can conclude that in order to pick-up the dynamics of the population any finite difference numerical scheme must select its step-size such that

$$\Delta t \ll T = \frac{1}{\lambda_1}. \tag{9.3.7}$$

Example B: Fisher equation

If diffusion is added to the logistic equation, then the resulting PDE is the Fisher equation

$$\frac{\partial u}{\partial t} = D\frac{\partial^2 u}{\partial x^2} + \lambda_1 u - \lambda_2 u^2, \quad u = u(x, t), \tag{9.3.8}$$

where D is the positive diffusion coefficient and it and the other variables have the physical units of

$$[t] = T, \quad [x] = L, \quad [u] = \#, \quad [D] = \frac{L^2}{T}, \tag{9.3.9}$$

where $[\lambda_1]$ and $[\lambda_2]$ are given in Eq. (9.3.2), and L is the physical unit of length.

Now transform to dimensionless variables $(\bar{v}, \bar{t}, \bar{x})$

$$u = U\bar{v}, \quad t = T\bar{t}, \quad x = X\bar{x}, \tag{9.3.10}$$

with scales (U, T, X). Substituting these items into Eq. (9.3.8) and simplifying gives

$$\frac{\partial \bar{v}}{\partial \bar{t}} = \left(\frac{DT}{X^2}\right)\frac{\partial^2 \bar{v}}{\partial \bar{x}^2} + (T\lambda_1)\bar{v} - (\lambda_2 UT)\bar{v}^2. \tag{9.3.11}$$

Setting the coefficients, on the right-side, to zero gives

$$\frac{DT}{X^2} = 1, \quad T\lambda_1 = 1, \quad \lambda_2 UT = 1, \tag{9.3.12}$$

and the scales

$$U = \frac{\lambda_1}{\lambda_2}, \quad T = \frac{1}{\lambda_1}, \quad X = \sqrt{\frac{D}{\lambda_1}}. \tag{9.3.13}$$

Thus, any finite difference discretization of the Fisher PDE must have the following restrictions on its two step-sizes

$$\Delta t \ll \frac{1}{\lambda_1}, \quad \Delta x \ll \sqrt{\frac{D}{\lambda_1}}, \tag{9.3.14}$$

to actually determine the variation of $u(x,t)$ as functions of x and t.

Example C: Combustion PDE

This PDE is

$$\frac{\partial u}{\partial t} = D\frac{\partial^2 u}{\partial t^2} + \lambda_3 u^2 - \lambda_4 u^3. \tag{9.3.15}$$

Proceeding as for the previous two examples, we obtain

$$u = U\bar{v}, \quad t = T\bar{t}, \quad x = X\bar{x}, \tag{9.3.16}$$

where

$$U = \frac{\lambda_1}{\lambda_2}, \quad T = \frac{\lambda_2}{\lambda_1^2}, \quad X = \sqrt{\frac{D\lambda_2}{\lambda_1^2}}. \tag{9.3.17}$$

The step-size restrictions are

$$\Delta t \ll \frac{\lambda_2}{\lambda_1^2}, \quad \Delta x \ll \sqrt{\frac{D\lambda_2}{\lambda_1^2}}. \tag{9.3.18}$$

Note that dropping the "bars" on the variables, the logistic ODE, the Fisher PDE, and the combustion PDE have the, respective, forms

$$\frac{du}{dt} = u - u^2, \tag{9.3.19a}$$

$$\frac{\partial u}{\partial t} = \frac{\partial^2 u}{\partial x^2} + u - u^2, \tag{9.3.19b}$$

$$\frac{\partial u}{\partial t} = \frac{\partial^2 u}{\partial x^2} + u^2 - u^3, \tag{9.3.19c}$$

i.e., the scaled differential equations contain no arbitrary numerical parameters.

9.4   Dynamical Consistency [2]

Dynamical consistency is applied with respect to particular features or properties of a system, and will generally change from one system to another. Mathematical models and their finite difference discretizations will therefore only be valid if the two are dynamically consistent with each other.

Definition: dynamic consistency

Consider a differential equations
$$\frac{dx}{dt} = f(x, t, \lambda), \quad x(0) = x_0, \tag{9.4.1}$$
where $\lambda = (\lambda_1, \lambda_2, \dots, \lambda_m)$ is the set of parameters characterizing the mathematical model. Let a discrete finite difference representation of Eq. (9.4.1) be

$$x_{k+1} = F(x_k, t_k, h, \lambda) \tag{9.4.2}$$

where $h = \Delta t$, $t_k = hk$, and x_k is an approximation to $x(t_k)$.

Let the differential equation and/or its solutions have property P. The discrete model, Eq. (9.4.2) is dynamically consistent (DC) with Eq. (9.4.1), with respect to the property P, if it and/or its solutions also have property P.

Examples of P include, but are not limited to the following items:

- positivity of solutions
- boundedness of solutions
- monotonicity of solutions
- number and stability of fixed-points
- overall order of the differential equation
- conservation laws
- special symmetries
- asymptotic behavior
- bifurcations

Note that this definition of DC can be easily extended to partial differential equations and systems of differential equations.

The purpose of introducing the concept of DC is to use it as a general principle to place restrictions on the mathematical forms which can arise in constructing valid NSFD schemes. (See Mickens [2] for a broad range of such applications.)

9.5 Numerical Instabilities

Numerical instabilities (NI) are solutions to the finite difference discretized equations, for a set of differential equations, which do not correspond to any solution of the original differential equations [3]. Their genesis lies in the fact that the parameter space of the finite difference scheme is always larger than the parameter space of the differential equation. This is easily seen for a single ODE

$$\frac{du}{dt} = g(u, t, \lambda), \tag{9.5.1}$$

where $\lambda = (\lambda_1, \lambda_2, \ldots, \lambda_m)$ are parameters appearing in the ODE. The corresponding discretizations take the form

$$u_{k+1} = G(u_k, t_k, \lambda, h), \tag{9.5.2}$$

where $h = \Delta t$ is the time step-size. Note that while the ODE has m-parameters, the finite difference scheme has $(m + 1)$ parameters. Since, in general, the only requirement on h is $h > 0$, the changing of h can cause the nature of the solutions to Eq. (9.5.2) to be modified as h varies. Thus, h can be a bifurcation parameter. This was shown to be the case in earlier sections of this book; see, for example, Chapter 2, and Mickens and Washington [4].

An elementary, but very powerful, explicit illustration of this type of bifurcation is the forward-Euler discretization of the decay equation

$$\frac{dx}{dt} = -\lambda x, \quad x(0) = x_0, \quad \lambda > 0, \tag{9.5.3}$$

which is

$$\frac{x_{k+1} - x_k}{h} = -\lambda x_k \rightarrow x_{k+1} = (1 - \lambda h) x_k. \tag{9.5.4}$$

It has several bifurcations, the first occurring at

$$h = \frac{1}{\lambda}. \tag{9.5.5}$$

The details are given in Section 2.2; see Mickens [5].

Finally, it should be indicated that for a central-difference scheme, i.e.,

$$\frac{x_{k+1} - x_{k-1}}{2h} = -\lambda x_k, \tag{9.5.6}$$

all the solutions are "chaotic" [6], i.e., they oscillate and become unbounded.

An essential purpose of the NSFD methodology is to insure that NI do not occur or if they do exist, they are rapidly damped.

9.6   Denominator Functions

A major feature of the NSFD methodology is concerned with how the first-order derivative should be discretized [2, 7, 8]. Within this framework, we express the discrete first-derivative as the form

$$\frac{dx(t)}{dt} \rightarrow \frac{x_{k+1} - \psi(h)x_k}{\phi(h)}, \qquad (9.6.1)$$

where, respectively, $\phi(h)$ and $\psi(h)$ are called the "denominator" and "numerator" functions. In general, both $\phi(h)$ and $\psi(h)$ depend on the time step-size, h, and have the mathematical properties

$$\psi(h) = 1 + O(h^2), \quad \phi(h) = h + O(h^2). \qquad (9.6.2)$$

For simplicity, we will use

$$\psi(h) = 1, \qquad (9.6.3)$$

and the task becomes evaluating or calculating $\phi(h)$ for a given system of ODE's.

The two papers of Mickens [7, 8] discuss a procedure for determining $\phi(h)$. For a single ODE, suppose it can be written in the form

$$\frac{dx}{dt} = ax + f_1(x), \qquad (9.6.4)$$

where $f_1(x)$ contains only nonlinear terms. Then, there are two possibilities for $\phi(h)$ and they are

$$\frac{x_{k+1} - x_k}{\phi_1(h,a)} = ax_k + (\dots)_k, \qquad (9.6.5a)$$

$$\phi_1(h,a) = \frac{e^{ah} - 1}{a}, \qquad (9.6.5b)$$

or

$$\frac{x_{k+1} - x_k}{\phi_2(h,a)} = ax_{k+1} + (\dots)_k, \qquad (9.6.6a)$$

$$\phi_2(h,a) = \frac{1 - e^{-ah}}{a}. \qquad (9.6.6b)$$

Note that a can be of either sign. Also, the notation, $(\dots)_k$, indicates a NSFD discretization for the nonlinear terms.

For the case where $a = 0$, i.e., no linear term occurs, then the denominator function is just h; therefore, we have

$$\phi(h,0) = h. \qquad (9.6.7)$$

An example of this is

$$\frac{dx}{dt} = bx^2, \tag{9.6.8}$$

where

$$\frac{x_{k+1} - x_k}{h} = (bx^2)_k = bx_{k+1}x_k. \tag{9.6.9}$$

These results can be easily extended to systems of coupled ODE's, especially those satisfying conservation laws [7, 8].

9.7 Nonlocal Discretization of Functions

If the discretization of a differential equation is to be dynamical consistent with respect to some proper $\bar{P}$ of the differential equation and/or its solutions, then it will, in general, require a nonlocal discretization of functions appearing in the differential equations. By nonlocal discretization, we mean that the factors in a particular term in the differential equation are to be evaluated at more than one computational lattice point. So, for example,

$$x \to 2x_k - x_{k+1}, \tag{9.7.1}$$

$$x^2 \to 3x_k^3 - x_k^2 x_{k+1}, \tag{9.7.2}$$

$$x^3 \to \left(\frac{x_{k+1} + x_{k-1}}{2}\right) x_k^2, \tag{9.7.3}$$

$$x^2 \to \left(\frac{x_{k+1} + x_k + x_{k-1}}{3}\right) x_k. \tag{9.7.4}$$

The above listing does not imply that whenever x appears, it should be discretely modeled by the expression of Eq. (9.7.1); it may just be required to use

$$x \to x_k. \tag{9.7.5}$$

For a given differential equation, it requires a deep and fundamental knowledge of the differential equation and the behavior of its solutions to select the appropriate nonlocal discretizations. But, in general, this knowledge is attainable through insight and experience.

Important point: The above discussion implies that there does not exist a "black-box NSFD scheme algorithm;" every equation has to be considered on its own terms. But, this is not really a difficulty as the work in Chapter 11 demonstrates.

The above considerations can be extended to PDE's. So, for example

$$u(x,t) \to \frac{u_{m+1}^k + u_m^k + u_{m-1}^k}{3}, \quad u_m^{k+1}, \tag{9.7.6}$$

$$u(x,t)^2 \to u_m^{k+1} u_{m-1}^k, \quad \left(\frac{u_{m+1}^k + u_{m-1}^k}{2}\right) u_m^k, \tag{9.7.7}$$

$$u(x,t)^3 \to \left(\frac{u_{m+1}^k + 2u_m^k + u_{m-1}^k}{3}\right)(u_m^k)^2. \tag{9.7.8}$$

9.8   Method of Sub-Equations

The method of sub-equations is a powerful tool to aid in the construction of NSFD schemes. A sub-equation is defined as follows:

> Let a differential equation consist of $N \geq 3$ terms. A sub-equation is any differential equation formed by including any $(N-1)$ or fewer terms of the original differential equation. (See Mickens [1], Section 9.7.3.)

To illustrate this concept, consider the Burgers PDE

$$u_t + uu_x = Du_{xx}. \tag{9.8.1}$$

There are three sub-equations

$$u_t = Du_{xx}, \tag{9.8.2}$$

$$u_t + uu_x = 0, \tag{9.8.3}$$

$$uu_x = Du_{xx}, \tag{9.8.4}$$

where the first two are PDE's, while the third is an ODE.

A second example is a linearized version of the Burgers equation

$$u_t + au_x = Du_{xx}, \tag{9.8.5}$$

where a and D are constants. The three sub-equations are

$$u_t + au_x = 0, \tag{9.8.6}$$

$$u_t = Du_{xx}, \tag{9.8.7}$$

$$au_x = Du_{xx}. \tag{9.8.8}$$

Note that exact finite difference schemes exist and are known for Eqs. (9.8.6) and (9.8.8). Also, observe that both of these differential equations contain the term, au_x. This means that a NSFD scheme for Eq. (9.8.5) can be constructed by combining the two exact discretizations in a way that the au_x term has the same discretization. The generalization of this idea often allows the construction of NSFD schemes.

9.9   Constructing NSFD Schemes

Below, we list a general set of steps that can be followed for the process of constructing valid NSFD schemes for differential equations. These steps are only suggestive of what should be done, as there does not exist a rigid, deterministic procedure. And, as we have stated several times earlier in this chapter, such exact rules of construction do not, even in principle, exist.

Thoughtful Rules for NSFD Scheme Constructions

(A) Start with the given set of differential equations and assure yourself of their "correctness."

(B) Examine the equations to see what parameters appear, their signs, possible magnitudes, etc.

(C) Determine the physical scales of the dependent and independent variables in terms of the parameters. For certain situations, the initial and/or boundary condition may play roles in the process.

(D) Consider the relevant sub-equations and determine the conditions under which their solutions provide insights into the behavior of the solutions to the full set of differential equations.

(E) Investigate, in as much detail as possible, the qualitative and quantitative properties of the differential equations.

(F) From the results in (E), determine a list of critical properties that you wish to have incorporated into the finite difference scheme.

(G) Calculate the denominator functions.

(H) Use insights, experience, and "tricks of the trade" to determine the nonlocal mathematical structure of the nonlinear terms in the differential equations.

(I) Check your discretization to verify that all of the previously selected properties have been incorporated, in a dynamically consistent manner, into the discretization.

(J) If necessary, repeat this cycle.

9.10   Final Comments

There is not a single NSFD model of a system of differential equations. The validity and numerical accuracy of a particular finite difference scheme are determined by the number of dynamical consistent properties built into the schemes. In general, we expect discretizations having more dynamical consistent features will be "better" than schemes having fewer such features.

Also if one is not familiar with the derivation of the equations, what purposes they and their solutions are to service, and lack insights into exactly what discretization means, then there is little chance of them successfully constructing valid NSFD schemes.

To repeat and summarize, the following comments conclude this chapter:

• The NSFD methodology is not based, in any of its aspects, on *a priori* knowledge of the exact analytical solutions to the differential equations.

• In general, the NSFD methodology does not generate exact finite difference schemes. However, the schemes they produce are "much better" than those obtained by the use of many of the standard techniques.

• The application of the principle of dynamical consistency places severe restrictions on the mathematical structures of both the denominator functions and the nonlocal forms of terms in the differential equations.

References

1. R. E. Mickens, *Difference Equations: Theory, Applications and Advanced Topics*, 3rd Edition (Chapman and Hall, Boca Raton, 2015).
2. R. E. Mickens, *Journal of Difference Equations and Applications* **11**, 645–653 (2005). Dynamic consistency: a fundamental principle for constructing NSFD schemes for differential equations.
3. B. F. Woods, *Numerical Instabilities in Finite-Difference Models of Differential Equations*, MS Thesis (Department of Mathematical Sciences, Atlanta University; Atlanta, GA; June 1989).
4. R. E. Mickens and T. M. Washington, *Abstracts of Papers Presented to the American Mathematics Society* (2019 Joint Mathematics Meetings; Baltimore, MD). Abstract 1145-39-1405. Bifurcations as the genesis of numerical instabilities in numerical solutions to differential equations.
5. R. E. Mickens, in I. W. Knowles and Y. Saitō (editors), *Differential Equations and Mathematical Physics* (Lecture Notes in Mathematics, Vol. 1285, Springer, Berlin, 1987). Runge-Kutta schemes and numerical instabilities: The logistic equation.
6. S. Ushiki, *Physics* **4D**, 407–424 (1982), Central difference scheme and chaos.
7. R. E. Mickens, *Numerical Methods for Partial Differential Equations* **32**, 672–691 (2007). Calculation of denominator functions

for NSFD schemes for differential equations satisfying a positivity condition.

8. R. E. Mickens, *Journal of Biological Dynamics* **1**, 427–436 (2007). Numerical integration of population models satisfying conservation laws: NSFD methods.

Chapter 10

Some Exact Finite Difference Schemes

10.1 Introduction

This chapter examines eight explicit examples of linear and nonlinear differential equations for which their exact finite difference representations can be calculated. Our interest in these equations is due to their appearance in the mathematical modeling of a broad range of phenomena in the natural and engineering sciences.

Detailed examination of these equations allows one to see the often clever algebraic manipulations required to achieve the desired exact discretizations.

10.2 General, Linear, Homogeneous, First-Order ODE

This differential equation is

$$\begin{cases} a(x)\frac{dy(x)}{dx} + b(x)y(x) = 0, \\ y(x_0) - y_0 \text{ (given)}, \end{cases} \tag{10.2.1}$$

and since it is separable, it can be easily integrated to give the solution

$$y(x) = y(x_0)\exp\left[-\int_{x_0}^{x} \frac{b(z)}{a(z)} \cdot dz\right]. \tag{10.2.2}$$

From this expression, it follows that

$$y(x + h) = y(x_0)\exp\left[-\int_{x_0}^{x+h}(\cdots)dz\right]$$

$$= y(x_0)\exp\left[-\int_{x_0}^{x}(\cdots)dz - \int_{x}^{x+h}(\cdots)dz\right]$$

$$= y(x)\exp\left[-\int_{x}^{x+h}(\cdots)dz\right], \tag{10.2.3}$$

where

$$(\cdots) = \frac{b(z)}{a(z)}. \tag{10.2.4}$$

Also

$$y(x+h) - y(x) = -y(x)\left[1 - e^{-c(x,h)}\right], \tag{10.2.5}$$

where

$$c(x,h) = \int_x^{x+h} (\cdots)dz. \tag{10.2.6}$$

Therefore, multiplying both sides of Eq. (10.2.5) by $[a(x+h)+a(x)]/2$, and rearranging the resultant expression, gives

$$\left[\frac{a(x+h) + a(x)}{2}\right]\left[\frac{y(x+h) - y(x)}{\phi(x,h)}\right] + b(x)y(x) = 0, \tag{10.2.7}$$

where

$$\phi(x,h) = \left[\frac{a(x+h) + a(x)}{2}\right]\left[\frac{1 - e^{-c(x,h)}}{b(x)}\right]. \tag{10.2.8}$$

The results given in Eqs. (10.2.7) and (10.2.8) provide an exact finite difference scheme for Eq. (10.2.1).

If we define

$$\begin{cases} x_k = hk, & y_k = y(x_k), \\ a_k = a(x_k), & b_k = b(x_k), \end{cases} \tag{10.2.9}$$

then the discretization becomes

$$\left(\frac{a_{k+1} + a_k}{2}\right)\left[\frac{y_{k+1} - y_k}{\phi}\right] + b_k y_k = 0, \tag{10.2.10}$$

with

$$\phi = \left(\frac{a_{k+1} + a_k}{2b_k}\right)\left\{1 - \exp\left[-\int_{x_k}^{x_{k+1}} \frac{b(z)}{a(z)} \cdot dz\right]\right\}. \tag{10.2.11}$$

Note for the special case where $a(x) = a_0$ and $b(x) = b_0$ are constant, the exact scheme reduces to the known results

$$a_0\left[\frac{y(x+h) - y(x)}{\phi}\right] + b_0 y(x) = 0, \tag{10.2.12a}$$

$$\phi = \frac{1 - e^{-\left(\frac{b_0}{a_0}\right)h}}{\left(\frac{a_0}{b_0}\right)}. \tag{10.2.12b}$$

10.3 Several Important Exact Schemes

There are certain differential equations which often appear as subequations in more complex equations. Therefore, it is of great value to have knowledge of their exact finite difference schemes.

The eleven differential equations given in this section and their corresponding finite difference discretizations have all been considered earlier in the book, particularly, Chapters 3, 6, 7. Thus, we collect them together mainly for purposes of convenience, i.e., ease of locating them for future use in the construction of NSFD schemes.

10.3.1 *Decay Equation*

$$\frac{dy}{dt} = -\lambda y, \tag{10.3.1}$$

$$\frac{y_{k+1} - y_k}{\left(\frac{1-e^{-\lambda h}}{\lambda}\right)} = -\lambda y_k \quad \text{or} \quad \frac{y_{k+1} - y_k}{\left(\frac{e^{\lambda h}-1}{\lambda}\right)} = -\lambda y_{k+1}. \tag{10.3.2}$$

10.3.2 *Harmonic Oscillator*

$$\frac{d^2 y}{dt^2} + \omega^2 y = 0, \tag{10.3.3}$$

$$\frac{y_{k+1} - 2y_k + y_{k-1}}{\left(\frac{4}{\omega^2}\right)\sin^2\left(\frac{h\omega}{2}\right)} + \omega^2 y_k = 0. \tag{10.3.4}$$

Note that for a negative sign in front of the ω^2 term, we have

$$\frac{d^2 y}{dt^2} - \omega^2 y = 0, \tag{10.3.5}$$

$$\frac{y_{k+1} - 2y_k + y_{k-1}}{\left(\frac{4}{\omega^2}\right)\sinh^2\left(\frac{h\omega}{2}\right)} - \omega^2 y_k. \tag{10.3.6}$$

10.3.3 *Logistic Equation*

$$\frac{dy}{dt} = \lambda_1 y - \lambda_2 y^2 \tag{10.3.7}$$

$$\frac{y_{k+1} - y_k}{\left(\frac{e^{\lambda_1 h}-1}{\lambda_1}\right)} = \lambda_1 y_k - \lambda_2 y_{k+1} y_k. \tag{10.3.8}$$

10.3.4   *Quadratic Decay Equation*

$$\frac{dy}{dt} = -y^2, \tag{10.3.9}$$

$$\frac{y_{k+1} - y_k}{h} = -y_{k+1}y_k. \tag{10.3.10}$$

10.3.5   *Nonlinear Equation*

$$2\frac{dy}{dt} + y = \frac{1}{y}, \tag{10.3.11}$$

$$2\left[\frac{y_{k+1} - y_k}{(1 - e^{-h})}\right] + \frac{y_k^2}{\left(\frac{y_{k+1}+y_k}{2}\right)} = \frac{1}{\left(\frac{y_{k+1}+y_k}{2}\right)}. \tag{10.3.12}$$

10.3.6   *Cubic Decay Equation*

$$\frac{dy}{dt} = -y^3, \tag{10.3.13}$$

$$\frac{y_{k+1} - y_k}{h} = -\left(\frac{2y_{k+1}}{y_{k+1} + y_k}\right) y_{k+1}y_k^2. \tag{10.3.14}$$

10.3.7   *Linear Velocity Force Equation*

$$\frac{d^2y}{dt^2} = \lambda \frac{dy}{dt}, \tag{10.3.15}$$

$$\frac{y_{k+1} - 2y_k + y_{k-1}}{\left(\frac{e^{\lambda h}-1}{\lambda}\right)h} = \lambda \left(\frac{y_k - y_{k-1}}{h}\right). \tag{10.3.16}$$

10.3.8   *Damped Harmonic Oscillator*

$$\frac{d^2y}{dt^2} + 2\epsilon\frac{dy}{dt} + y = 0, \tag{10.3.17}$$

$$\left[\frac{y_{k+1} - 2y_k + y_{k-1}}{\phi^2}\right] + 2\epsilon\left(\frac{y_k - \psi y_{k-1}}{\phi}\right)$$
$$+ \left[\frac{2(1 - \psi)y_k + (\phi^2 + \psi^2 - 1)y_{k-1}}{\phi^2}\right] = 0, \tag{10.3.18}$$

where

$$\phi(\epsilon, h) = \frac{e^{-\epsilon h}}{\sqrt{1 - \epsilon^2}} \cdot \sin\left(\sqrt{1 - \epsilon^2}\right) h \tag{10.3.19}$$

$$\psi(\epsilon, h) = \frac{\epsilon e^{-\epsilon h}}{\sqrt{1 - \epsilon^2}} + e^{-\epsilon h} \cos\left(\sqrt{1 - \epsilon^2}\right) h. \tag{10.3.20}$$

10.3.9 *Unidirectional Wave Equations*

$$u_t + u_x = 0, \quad u = u(x,t), \tag{10.3.21}$$

$$\frac{u_m^{k+1} - u_m^k}{\phi(\Delta t)} + \frac{u_m^k - u_{m-1}^k}{\phi(\Delta x)} = 0, \quad \Delta t = \Delta x, \tag{10.3.22}$$

and

$$u_t - u_x = 0, \quad u = u(x,t), \tag{10.3.23}$$

$$\frac{u_m^{k+1} - u_m^k}{\psi(\Delta t)} - \frac{u_{m+1}^k - u_m^k}{\psi(\Delta x)} = 0, \quad \Delta t = \Delta x, \tag{10.3.24}$$

where

$$\phi(z) = z + O(z^2), \quad \psi(z) = z + O(z^2), \tag{10.3.25}$$

otherwise $\phi(z)$ and $\psi(z)$ are arbitrary.

10.3.10 *Full Wave Equation*

$$u_{tt} = u_{xx}, \quad u = u(x,t), \tag{10.3.26}$$

$$\frac{u_m^{k+1} - 2u_m^k + u_m^{k-1}}{\phi(\Delta t)} + \frac{u_{m+1}^k - 2u_m^k + u_{m-1}^k}{\phi(\Delta x)}, \tag{10.3.27}$$

with

$$\Delta t = \Delta x, \quad \phi(z) = z + O(z^2). \tag{10.3.28}$$

10.3.11 *Nonlinear, Fisher-Type Unidirection Wave Equation*

$$u_t + u_x = u(1 - u), \quad u = u(x,t), \tag{10.3.29}$$

$$\frac{u_m^{k+1} - u_m^k}{(e^{\Delta t} - 1)} + \frac{u_m^k - u_{m-1}^k}{(e^{\Delta x} - 1)} = u_{m-1}^k(1 - u_m^{k+1}), \tag{10.3.30}$$

$$\Delta t = \Delta x. \tag{10.3.31}$$

10.3.12　*Unidirectional, Spherical Wave Equation* [1]

$$\frac{\partial u}{\partial t} + \frac{u}{r} + \frac{\partial u}{\partial r} = 0, \quad u = u(r,t), \tag{10.3.32}$$

$$\frac{u^{k+1} - u_m^k}{\Delta t} + \frac{u_{m-1}^k}{r_m} + \frac{u_m^k - u_{m-1}^k}{\Delta r} = 0, \tag{10.3.33}$$

with

$$\begin{cases} t_k = (\Delta t)k, \quad x_m = (\Delta r)m \\ u(r_m, t_k) = u_m^k \end{cases} \tag{10.3.34}$$

and

$$\Delta t = \Delta r. \tag{10.3.35}$$

10.3.13　*Steady-State Wave Equation with Spherical Symmetry* [2]

$$\frac{d^2\psi}{dr^2} + \left(\frac{2}{r}\right)\frac{d\psi}{dr} + k^2\psi = 0, \quad \psi = \psi(r), \tag{10.3.36}$$

$$\frac{\psi_{m+1} - 2\psi_m + \psi_{m-1}}{\left(\frac{4}{k^2}\right)\sin^2\left(\frac{k\Delta r}{2}\right)} + \left(\frac{2}{r_m}\right)\frac{\psi_{m+1} - \psi_{m-1}}{\left(\frac{2}{\Delta r}\right)\left(\frac{4}{k^2}\right)\sin^2\left(\frac{k\Delta r}{2}\right)} + k^2\psi_m = 0. \tag{10.3.37}$$

10.3.14　*Wave Equation having Spherical Symmetry* [2]

$$\frac{\partial^2\phi}{\partial r^2} + \left(\frac{2}{r}\right)\frac{\partial\phi}{\partial r} = \frac{1}{c^2}\frac{\partial^2\phi}{\partial t^2}, \quad \phi = \phi(r,t), \tag{10.3.38}$$

$$\frac{\phi_{m+1}^k - 2\phi_m^k + \phi_{m-1}^k}{(\Delta r)^2} + \left(\frac{2}{r_m}\right)\left(\frac{\phi_{m+1}^k - \phi_{m-1}^k}{2\Delta r}\right)$$
$$= \left(\frac{1}{c^2}\right)\left(\frac{\phi_m^{k+1} - 2\phi_m^k + \phi_m^{k-1}}{(\Delta t)^2}\right) \tag{10.3.39}$$

with

$$\Delta r = c\Delta t \tag{10.3.40}$$

and

$$\phi_m^k = \phi(r_m, t_k). \tag{10.3.41}$$

10.3.15 *Two-Dimensional, Linear Advection Equation* [3]

$$u_t + au_x + bu_y = 0, \quad u = u(x, y, t), \tag{10.3.42}$$

$$\frac{u_{m,n}^{k+1} - u_{m,n}^k}{\Delta t} + \left(\frac{a}{\Delta x}\right)\left[\frac{u_{m,n}^k + u_{m,n-1}^k}{2} - \frac{u_{m-1,n}^k + u_{m-1,n-1}^k}{2}\right]$$

$$+ \left(\frac{b}{\Delta y}\right)\left[\frac{u_{m-1,n}^k + u_{m,n}^k}{2} - \frac{u_{m-1,n}^k + u_{m,n-1}^k}{2}\right] = 0, \tag{10.3.43}$$

with

$$\Delta x = b\Delta t, \quad \Delta y = b\Delta t, \tag{10.3.44}$$

and

$$u_{m,n}^k = u(x_m, y_n, t_k). \tag{10.3.45}$$

10.3.16 *Two-Dimensional, Nonlinear (Logistic) Advection Equation* [3]

$$u_t + au_x + bu_y = u(1 - u), \quad u = u(x, y, t), \tag{10.3.46}$$

$$\frac{u_{m,n}^{k+1} - u_{m,n}^k}{\phi_1(\Delta t)} + a\left[\frac{u_{m,n}^k - u_{m-1,n}^k}{\phi_2(\Delta x)}\right]$$

$$+ b\left[\frac{u_{m-1,n}^k - u_{m-1,n-1}^k}{\phi_3(\Delta y)}\right]$$

$$= u_{m-1,n-1}^k(1 - u_{m,n}^{k+1}), \tag{10.3.47}$$

where

$$\Delta x = a\Delta t, \quad \Delta y = b\Delta t, \tag{10.3.48}$$

$$u_{m,n}^k = u(x_m, y_n, t_k), \tag{10.3.49}$$

and

$$\phi_1(\Delta t) = e^{\Delta t} - 1, \quad \phi_2(\Delta x) = a\left[e^{\frac{\Delta x}{a}} - 1\right], \quad \phi_3(\Delta y) = b\left[e^{\frac{\Delta y}{b}} - 1\right]. \tag{10.3.50}$$

10.4   Two Coupled, Linear ODE's with Constant Coefficients

Consider the following system of two coupled, linear differential equations with constant coefficients

$$\frac{dx}{dt} = ax + by, \tag{10.4.1a}$$

$$\frac{dy}{dt} = cx + dy. \tag{10.4.1b}$$

Section 6.3 of this book, provides the details for the construction of the corresponding exact finite difference scheme; it is

$$\frac{x_{k+1} - \psi x_k}{\phi} = ax_k + by_k, \tag{10.4.2a}$$

$$\frac{y_{k+1} - \psi y_k}{\phi} = cx_k + dy_k, \tag{10.4.2b}$$

where

$$\psi = \frac{\lambda_1 e^{\lambda_2 h} - \lambda_2 e^{\lambda_1 h}}{\lambda_1 - \lambda_2}, \tag{10.4.3a}$$

$$\phi = \frac{e^{\lambda_1 h} - e^{\lambda_2 h}}{\lambda_1 - \lambda_2}, \tag{10.4.3b}$$

and λ_1 and λ_2 are the roots of the characteristic equation [4]

$$\det \begin{bmatrix} a - \lambda & b \\ c & d - \lambda \end{bmatrix} = 0. \tag{10.4.4}$$

Note that ψ and ϕ have the properties

$$\psi = 1 + O(h^2), \quad \phi = h + O(h^2). \tag{10.4.5}$$

Further, observe that the right-sides of Eqs. (10.4.2), the variables (x, y) are evaluated at $t = t_k$.

A generalization of the above result was made by my collaborator, Professor Lih-Ing W. Roeger [5], by demonstrating that another possibility exists for constructing exact finite difference schemes. Briefly, this method begins by writing Eqs. (10.4.1) in matrix form, i.e.,

$$\mathbf{x} = \begin{pmatrix} x \\ y \end{pmatrix}, \quad \mathbf{A} = \begin{pmatrix} a & b \\ c & d \end{pmatrix}, \tag{10.4.6}$$

and then demanding that the following expression is an exact finite difference scheme

$$\frac{\mathbf{x}_{k+1} - \mathbf{x}_k}{\phi} = \mathbf{A} \left[\theta \mathbf{x}_{k+1} + (1 - \theta) \mathbf{x}_k \right]. \tag{10.4.7}$$

This requirement allows for the explicit calculation of the scalar functions (ϕ, θ) in terms of h and (λ_1, λ_2). In terms of the Jordan form matrix of A, Roeger lists in Table 1, of her publication, explicit expressions for $(\mathbf{J}, \phi, \theta)$; there are ten cases.

A recent paper by Quang and Tuan [6] extends the results obtained by Roeger to the case of three coupled, linear ODE's having constant coefficients. For this case

$$\frac{d\mathbf{x}(t)}{dt} = \mathbf{A}\mathbf{x}(t), \quad \mathbf{x}(t) = \begin{pmatrix} x(t) \\ y(t) \\ z(t) \end{pmatrix}, \tag{10.4.8}$$

and they assume that the exact finite difference scheme takes the form

$$\frac{\mathbf{x}_{k+1} - \psi\mathbf{x}_k}{\phi} = \mathbf{A}[\theta\mathbf{x}_{k+1} + (1 - \theta)\mathbf{x}_k]. \tag{10.4.9}$$

Consequently, there are three scalar functions of h and the three eigenvalues of $\mathbf{A}$ to determine. Clearly, the algebraic details are more complicated than for the two-dimensional case.

10.5 Jacobi Cosine and Sine Functions

The Jacobi elliptic functions provide solutions for the following nonlinear, second-order differential equation

$$\frac{d^2y}{dt^2} + ay + by^3 = 0, \tag{10.5.1}$$

where the parameters (a, b) may be of either sign [7, 8]. The initial conditions

$$y(0) = A, \quad \frac{dy(0)}{dt} = 0; \quad a > 0, \quad b > 0; \tag{10.5.2}$$

$$y(0) = 0, \quad \frac{dy(0)}{dt} = A; \quad a > 0, \quad b < 0; \tag{10.5.3}$$

respectively, give solutions proportional to the Jacobi cosine and sine functions. An extensive survey of many of the applications for these functions appear in the books by Kovacic and Brennan [7], and McLachlan [8]. Also, an extensive listing of the properties of the functions are given in McLachlan et al. [8].

The standard notations for these functions are

$$\begin{cases} cn(t, k) : \text{Jacobi cosine} \\ sn(t, k) : \text{Jacobi sine} \end{cases}$$

where the parameter k is expressible in terms of the (a, b) of Eq. (10.5.1) [10].

In a "normalized" representation, the Jacobi cosine function, $cn(t, k)$, satisfies the differential equation

$$\begin{cases} \frac{d^2x}{dt^2} + (1 - 2k^2)x + 2k^2x^3 = 0, \\ x(0) = 1, \quad \frac{dx(0)}{dt} = 0, \end{cases} \tag{10.5.4}$$

where

$$x(t) = cn(t, k), \tag{10.5.5}$$

with

$$0 \leq k^2 < \frac{1}{2}. \tag{10.5.6}$$

It has been shown by Mickens and Washington [10] that the exact finite difference scheme for Eq. (10.5.4) is

$$\frac{x_{m+1} - 2x_m + x_{m-1}}{(sn\ h)^2} + 2\left\{\left[\frac{1 - (cn\ h)}{(sn\ h)^2}\right]x_m - k^2\left(\frac{x_{m+1} + x_{m-1}}{2}\right)\right\}$$
$$+ 2k^2\left(\frac{x_{m+1} + x_{m-1}}{2}\right)x_m^2 = 0, \tag{10.5.7}$$

where

$$x_m = x(t_m) = cn(t_m, k) \tag{10.5.8}$$

and

$$sn\ h = sn(h, k), \quad cn\ h = cn(h, k). \tag{10.5.9}$$

In a similar manner, the nonlinear differential equation for the Jacobi sine function, $x(t) = sn(t, k)$, and its exact finite difference scheme are, respectively, given by the expressions

$$\begin{cases} \frac{d^2x}{dt^2} + (1 + k^2)x - 2k^2x^3 = 0, \\ x(0) = 0, \quad \frac{dx(0)}{dt} = 1, \end{cases} \tag{10.5.10}$$

where

$$x(t) = sn(t, k), \tag{10.5.11}$$

with

$$0 \leq k^2 < 1, \tag{10.5.12}$$

and

$$\frac{x_{m+1} - 2x_m + x_{m-1}}{(sn\ h)^2} + 2\left[\frac{1 - (cn\ h)(dn\ h)}{(sn\ h)^2}\right] x_m$$
$$- 2k^2\left(\frac{x_{m+1} + x_{m-1}}{2}\right) x_m^2 = 0, \tag{10.5.13}$$

where

$$x_m = x(t_m) = sn(t_m, k). \tag{10.5.14}$$

The above derivations depend on using the two addition theorems [9]

$$cn(u + v) = \frac{(cn\ u)(cn\ v) - (sn\ u)(dn\ u)(sn\ v)(dn\ v)}{1 - k^2(sn\ u)^2(sn\ v)^2}, \tag{10.5.15}$$

$$sn(u + v) = \frac{(sn\ u)(cn\ v)(dn\ v) + (sn\ v)(cn\ u)(dn\ u)}{1 - k^2(sn\ u)^2(sn\ v)^2}, \tag{10.5.16}$$

along with the relations

$$(dn\ u)^2 = 1 - k^2(sn\ u)^2 \tag{10.5.17a}$$
$$(cn\ u)^2 + (sn\ u)^2 = 1. \tag{10.5.17b}$$

10.6 Cauchy-Euler Equation

The following second-order, linear ordinary differential equation

$$Dx^2\frac{d^2y}{dx^2} + rx\frac{dy}{dx} - ry = 0, \tag{10.6.1}$$

where D and r are positive constants, is a specific example of the Cauchy-Euler equation [11]

$$a_0x^2\frac{d^2y}{dx^2} + a_1x\frac{dy}{dx} + a_2y = 0, \tag{10.6.2}$$

where (a_0, a_1, a_2) are constants. The representation, given by Eq. (10.6.1) is the time independent part of what is called a Black-Scholes type partial differential equation (BSTPDE) [12, 13]

$$\frac{\partial U}{\partial t} = Dz^2\frac{\partial^2 U}{\partial z^2} + rz\frac{\partial U}{\partial z} - rU, \quad U = U(z,t). \tag{10.6.3}$$

In the following, we merely give the exact finite difference discretization of Eq. (10.6.1). The complete details of this derivation is presented in the paper by Mickens et al. [13]; see also the article by Roeger and Mickens [14].

Define $(x_m, y_m, A_m, B_m, C_m, \delta)$ as follows

$$x_m = (\Delta x)m, \quad y_m = y(x_m), \quad \delta = \frac{r}{D}, \tag{10.6.4}$$

$$A_m = \left(\frac{m+1}{m^\delta}\right) - \left[\frac{m}{(m+1)^\delta}\right], \tag{10.6.5a}$$

$$B_m = \left(\frac{m+2}{m^\delta}\right) - \left[\frac{m}{(m+2)^\delta}\right], \tag{10.6.5b}$$

$$C_m = \left(\frac{m+2}{(m+1)^\delta}\right) - \left[\frac{m+1}{(m+2)^\delta}\right]. \tag{10.6.5c}$$

It can be shown that

$$A_m > 0, \quad B_m > 0, \quad C_m > 0. \tag{10.6.6}$$

With this notation, the exact discretization of Eq. (10.6.1) is [13]

$$D\left[\left(\frac{r}{D}\right)\left(\frac{A_{m-1}}{B_{m-1} - A_{m-1} - C_{m-1}}\right)\right](y_{m+1} - 2y_m + y_{m-1})$$
$$+ r\left[\frac{2A_{m-1} - B_{m-1}}{B_{m-1} - A_{m-1} - C_{m-1}}\right](y_m - y_{m-1})$$
$$- ry_{m-1} = 0. \tag{10.6.7}$$

10.7   Michaelis-Menten Equation

The Michaelis-Menten difference equation (MMDE) is

$$\frac{dy}{dt} = -\frac{ay}{1 + by}, \quad y(t_0) = y_0 > 0, \tag{10.7.1}$$

with (a, b) non-negative. This equation has been used to model a broad range of phenomena in the natural sciences [15, 16, 17].

Our goal is to derive an exact finite difference discretization for the MMDE. To do this, we outline the procedure given in Mickens [18].

First, note that Eq. (10.7.1) is a separable first-order ODE and, as a consequence, integrates to the following implicit solution

$$\text{Ln}\, y + by + at = \text{Ln}\, y_0 + by_0 + at_0. \tag{10.7.2}$$

Second, an explicit solution to Eq. (10.7.2) can be expressed in terms of the Lambert W-function. This function is defined as the solution to the following transcendental equation [19]

$$W(t)e^{W(t)} = t. \tag{10.7.3}$$

Therefore, for the equation

$$\text{Ln}(A + By) + Cy = \text{Ln } D, \tag{10.7.4}$$

the solution is [19]

$$y = \left(\frac{1}{C}\right) W\left[\left(\frac{CD}{B}\right)\exp\left(\frac{AC}{B}\right)\right] - \left(\frac{A}{B}\right). \tag{10.7.5}$$

A comparison of Eqs. (10.7.2) and (10.7.4) gives

$$A = 0, \quad B = 1, \quad C = b, \tag{10.7.6a}$$

$$\text{Ln } D = \text{Ln } y_0 + by_0 + a(t_0 - t). \tag{10.7.6b}$$

Using the fact that D can be expressed as

$$D = y_0 \exp[by_0 - a(t - t_0)], \tag{10.7.7}$$

we can write $y(t)$ as

$$y(t) = \left(\frac{1}{b}\right) W[(by_0 e^{by_0})e^{-a(t-t_0)}]. \tag{10.7.8}$$

As a consistency check on the derivation, note that for $b \to 0$, Eq. (10.7.1) becomes

$$\frac{dy}{dt} = -ay, \quad y(t_0) = y_0. \tag{10.7.9}$$

Also, since for small t, i.e., $|t| \ll 1$, [19]

$$W(t) = t + O(t^2), \tag{10.7.10}$$

it follows that $y(t)$ in Eq. (10.7.8) has the property

$$\underset{b \to 0}{\text{Lim}}\, y(t) = y_0 e^{-a(t-t_0)}, \tag{10.7.11}$$

and this is the correct limiting solution for Eq. (10.7.1), as seen from the inspection of Eq. (10.7.9).

To construct an exact finite difference scheme for the MMDE, we make sure of the fact that if the solution to

$$\frac{dy}{dt} = f(y), \quad y(t_0) = y_0, \tag{10.7.12}$$

is

$$y(t) = \phi(y_0, t_0, t), \tag{10.7.13}$$

then the exact scheme is [20]

$$y_{k+1} = \phi(y_k, t_k, t_{k+1}). \tag{10.7.14}$$

Using this result, it follows that the exact finite difference discretization of the MMDE, Eq. (10.7.1), is

$$y_{k+1} = \left(\frac{1}{b}\right) W\left[(be^{-ah})y_k e^{by_k}\right], \tag{10.7.15}$$

and this can be rewritten as

$$\frac{y_{k+1} - y_k}{\phi} = \frac{W(\cdots) - by_k}{b\phi}, \tag{10.7.16}$$

where ϕ is, for the time being, an unknown denominator function [21], possibly depending on the parameters a and b, and having the property

$$\phi(h, a, b) = h + O(h^2). \tag{10.7.17}$$

Since Eq. (10.7.1) can be written

$$\frac{dy}{dt} = -ay + O(y^2), \tag{10.7.18}$$

then following the results provided in Mickens [21], the denominator function ϕ is given by the following expression

$$\phi = \phi(a, h) = \frac{1 - e^{-ah}}{a}. \tag{10.7.19}$$

10.8   Weierstrass Elliptic Function

The Weierstrass elliptic function, $\mathscr{P}(z)$, satisfies the following first-order, nonlinear, differential equation of second degree [22, 23]

$$\left[\frac{d\mathscr{P}(z)}{dz}\right]^2 = 4[\mathscr{P}(z)]^3 - g_2 \mathscr{P}(z) - g_3, \tag{10.8.1}$$

where the constants g_2 and g_3 are the so-called "invariants" [22, 23]. Note that taking the derivative of Eq. (10.8.1) and cancelling the common factor of $d\mathscr{P}(z)/dz$ gives

$$\frac{d^2 \mathscr{P}(z)}{dz^2} = 6[\mathscr{P}(z)]^2 - \left(\frac{1}{2}\right) g_2. \tag{10.8.2}$$

This function also satisfies an addition theorem [21]

$$\mathscr{P}(z+y) = \left(\frac{1}{4}\right) \left[\frac{\mathscr{P}'(z) - \mathscr{P}'(y)}{\mathscr{P}(z) - \mathscr{P}(y)}\right]^2 - \mathscr{P}(z) - \mathscr{P}(y), \qquad (10.8.3)$$

and the duplication formula

$$\mathscr{P}(2z) = \left(\frac{1}{4}\right) \left[\frac{\mathscr{P}''(z)}{\mathscr{P}'(z)}\right]^2 - 2\mathscr{P}(z). \qquad (10.8.4)$$

It has been shown by Potts [24] that exact finite difference representations can be constructed for both Eq. (10.8.1) and Eq. (10.8.2). They are, respectively,

$$\frac{(\mathscr{P}_{n+1} - \mathscr{P}_n)^2}{\left(\frac{1}{k}\right)} = 4\mathscr{P}_n \mathscr{P}_{n+1} \left(\frac{\mathscr{P}_n + \mathscr{P}_{n+1}}{2}\right)$$

$$- g_2 \left(\frac{\mathscr{P}_n + \mathscr{P}_{n+1}}{2}\right) - g_3$$

$$- \left(\frac{1}{k}\right) \left\{ \left[\mathscr{P}_n \mathscr{P}_{n+1} + \left(\frac{1}{4}\right) g_2\right]^2 + g_3(\mathscr{P}_n + \mathscr{P}_{n+1}) \right\}, \qquad (10.8.5)$$

and

$$\frac{\mathscr{P}_{n+1} - 2\mathscr{P}_n + \mathscr{P}_{n-1}}{\left(\frac{1}{k}\right)} = 6\mathscr{P}_n \left(\frac{\mathscr{P}_{n+1} + \mathscr{P}_n + \mathscr{P}_{n-1}}{3}\right) - \left(\frac{1}{2}\right) g_2$$

$$- \left(\frac{1}{k}\right) \left[\mathscr{P}_n^2(\mathscr{P}_{n+1} + \mathscr{P}_{n-1}) + \left(\frac{1}{2}\right) g_2 \mathscr{P}_n + g_3\right]. \qquad (10.8.6)$$

For these expressions, the following notation is used

$$z \to z_n = z_n = z_0 + hn, \quad h = \Delta z; \qquad (10.8.7)$$

$$\mathscr{P}(z) \to \mathscr{P}(z_n) = \mathscr{P}_n; \qquad (10.8.8)$$

$$k = \mathscr{P}(h) = \frac{1}{h^2} + O(h^2). \qquad (10.8.9)$$

Inspection of Eqs. (10.8.5) and (10.8.6) shows that the denominator function is

$$\phi(h) = \left[\frac{1}{\mathscr{P}(h)}\right]^{1/2}. \qquad (10.8.10)$$

Also, both the first-order and second-order discrete representations have an additional term of order h^2. These terms are, respectively, the fourth and third expressions in Eqs. (10.8.5) and (10.8.6).

10.9   Modified Lotka-Volterra Equations

The standard Lotka-Volterra (L-V) equations are [15, 17]

$$\frac{dx}{dt} = r_1 x - a_{11} x^2 - a_{12} xy \qquad (10.9.1a)$$

$$\frac{dy}{dt} = r_2 y - a_{21} xy - a_{22} y^2 \qquad (10.9.1b)$$

$$x(0) = x_0 > 0, \quad y(0) = y_0 > 0, \qquad (10.9.1c)$$

where $(r_1, r_2, a_{11}, a_{12}, a_{21})$ are constant parameters, and $x(t)$ and $y(t)$ are populations which can interact with each other.

If, on the right-sides, we make the replacements

$$x \to \sqrt{x}, \quad y \to \sqrt{y}, \qquad (10.9.2)$$

then the following new ODE's are obtained

$$\frac{dx}{dt} = r_1 \sqrt{x} - a_{11} x - a_{12} \sqrt{x} \sqrt{y}, \qquad (10.9.3a)$$

$$\frac{dy}{dt} = r_2 \sqrt{y} - a_{21} \sqrt{x} \sqrt{y} - a_{22} y. \qquad (10.9.3b)$$

Now, making the nonlinear transformation of variables

$$u = \sqrt{x}, \quad v = \sqrt{y}, \qquad (10.9.4)$$

and dividing by either $2u$ or $2v$ gives

$$\frac{du}{dt} = \bar{r}_1 - \bar{a}_{11} u - \bar{a}_{12} v, \qquad (10.9.5a)$$

$$\frac{dv}{dt} = \bar{r}_2 - \bar{a}_{21} u - \bar{a}_{22} v, \qquad (10.9.5b)$$

where all the barred parameters are the original parameters divided by two, i.e., $\bar{r}_1 = r_1/2$, $\bar{r}_2 = r_2/2$, etc.

Note that Eqs. (10.9.5) are two coupled, linear first-order differential equations, with all coefficients constant. Consequently, their exact solutions can be calculated [11]. Further, the exact finite difference schemes can also be determined; see Sections 3.2 and 6.3.

After a number of algebraic manipulations, Mickens [25] arrives at the following final expressions for the exact finite difference discretizations for the modified Lotka-Volterra equations, as presented in Eqs. (10.9.3),

$$\frac{x_{k+1} - x_k}{\phi} = \left[\left(a_{11} - \frac{2\psi_1}{\phi}\right) u^* + a_{12} v^*\right] \left(\frac{\sqrt{x_{k+1}} + \sqrt{x_k}}{2}\right)$$
$$+ \left(\frac{2\psi_1}{\phi} - a_{11}\right) \sqrt{x_k} \left(\frac{\sqrt{x_{k+1}} + \sqrt{x_k}}{2}\right)$$
$$- a_{12} \left(\frac{\sqrt{x_{k+1}} + \sqrt{x_k}}{2}\right) \sqrt{y_k}, \qquad (10.9.6a)$$

$$\frac{y_{k+1} - y_k}{\phi} = \left[\left(a_{22} - \frac{2\psi_1}{\phi}\right) v^* + a_{21} u^*\right] \left(\frac{\sqrt{y_{k+1}} + \sqrt{y_k}}{2}\right)$$
$$- a_{21}\sqrt{x_k} \left(\frac{\sqrt{y_{k+1}} + \sqrt{y_k}}{2}\right)$$
$$+ \left(\frac{2\psi_1}{\phi} - a_{22}\right) \left(\frac{\sqrt{y_{k+1}} + \sqrt{y_k}}{2}\right) \sqrt{y_k}. \qquad (10.9.6b)$$

In these expressions, the following definitions apply:

$$4\lambda_{1,2} = -(a_{11} + a_{22}) \pm \sqrt{(a_{11} + a_{22})^2 - 4(a_{11}a_{22} - a_{12}a_{21})} \qquad (10.9.7a)$$

$$\phi = \frac{e^{\lambda_1 h} - e^{\lambda_2 h}}{\lambda_1 - \lambda_2} \qquad (10.9.7b)$$

$$\psi = 1 + \psi_1 = \frac{\lambda_1 e^{\lambda_2 h} - \lambda_2 e^{\lambda_1 h}}{\lambda_1 - \lambda_2} \qquad (10.9.7c)$$

$$u^* = \frac{a_{12}r_2 - a_{22}r_1}{a_{12}a_{21} - a_{11}a_{22}}, \qquad v^* = \frac{a_{21}r_1 - a_{11}r_2}{a_{12}a_{21} - a_{11}a_{22}}. \qquad (10.9.7d)$$

Finally, it should be noted that in spite of the apparent complexity of the exact schemes for $x(t)$ and $y(t)$, if one goes to the exact scheme for the $u = \sqrt{x}$ and $v = \sqrt{y}$ variables, they are explicit, i.e.,

$$u_{k+1} = A_1 + B_1 u_k + C_1 v_k, \qquad (10.9.8a)$$

$$v_{k+1} = A_2 + B_2 u_k + C_2 v_k, \qquad (10.9.8b)$$

where $(A_1, A_2, B_1, B_2, C_1, C_2)$ are known; see Eqs. (4.4) in Mickens [25].

10.10   Comments

It is important to again remind ourselves why we are interested in having explicit, exact finite difference representations of a carefully chosen set of differential equations. This is because such equations often appear as sub-equations in the expressions of more complex differential equations.

Consequently, knowledge of their discretizations can aid in the construction of valid NSFD schemes for the complex equations. This technique was used for many of the examples presented in Chapters 3 to 7 and will be further illustrated in the work given in Chapter 11.

References

1. R. E. Mickens, *Journal of Sound and Vibration* **182**, 342–344 (1995). Exact finite difference scheme for a spherical wave equation.
2. R. E. Mickens, *Journal of Difference Equations and Applications* **2**, 263–269 (1996). Exact finite difference schemes for the wave equation with spherical symmetry.
3. R. E. Mickens, *Journal of Sound and Vibration* **207**, 426–428. Exact finite difference schemes for two-dimensional advection equation.
4. M. Sever, *Ordinary Differential Equations* (Boole Press, Dublin, 1987).
5. L.-I. Roeger, *Journal of Computational and Applied Mathematics* **219**, 102–109 (2008). Exact finite-difference schemes for two-dimensional linear systems with constant coefficients.
6. A. D. Quang and H. M. Tuan, *Vietnam Journal of Mathematics* **46**, 471–492 (2018). Exact finite difference schemes for three-dimensional linear systems with constant coefficients.
7. I. Kovacic and M. J. Brennan, *The Duffing Equation: Nonlinear Oscillators and Their Behaviors* (Wiley, New York, 2011).
8. N. W. McLachlan, *Ordinary Non-Linear Differential Equations in Engineering and Physical Sciences*, 2nd edition (Oxford University Press, New York, 1958).
9. F. W. J. Olver, D. W. Lozier, R. F. Bossvert, and C. W. Clark, *NIST Handbook of Mathematical Functions* (Cambridge University Press, New York, 2010). See Sections 22.6(i), 22.8(i), 22.10(i), and 22.13(iii).
10. R. E. Mickens and T. M. Washington, *Journal of Difference Equations and Applications* **19**, 1042–1047 (2013). A note on exact finite difference schemes for the differential equations satisfied by the Jacobi cosine and sine functions.
11. S. L. Ross, *Differential Equations* (Blairdell, Waltham, MA, 1964).
12. P. Wilmott, S. Howison, and J. Dewynne, *The Mathematics of Financial Derivatives* (Cambridge University Press, New York, 2002).

13. R. E. Mickens, J. Munyakazi, and T. M. Washington, *Journal of Difference Equations and Applications* **21**, 547–562 (2015) (doi.org/10.1080/10236198.2015.1034118). A note on the exact discretizatins for a Cauchy-Euler equation: Application to the Black-Scholes equation.

14. L.-I. W. Roeger and R. E. Mickens, *Journal of Difference Equations and Applications*. (doi:10.1080/10236198.2013.771635). Exact finite difference for linear differential equations with constant coefficients.

15. L. Edelstein-Keshet, *Mathematical Models in Biology* (McGraw-Hill, New York, 1976).

16. R. M. Anderson and R. M. May, *Population Biology of Infectious Diseases* (Springer-Verlag, Berlin, 1982).

17. J. D. Murray, *Mathematical Biology* (Springer-Verlag, Berlin, 1989).

18. R. E. Mickens, *Journal of Biological Dynamics* **5**, 391–397 (2011). (doi:10.1080/17513758.2010.515690). An exact discretization of Michaelis-Menten type population equations.

19. R. M. Corless, G. H. Gonnet, D. E. G. Hare, D. J. Jeffrey, and D. E. Knuth, *Advances in Computing and Mathematics* **5**, 329–359 (1996).

20. R. E. Mickens, *Nonstandard Finite Difference Models of Differential Equations* (World Scientific London, 1994).

21. R. E. Mickens, *Numerical Methods for Partial Differential Equations* **23**, 672–691 (2007). Calculation of denominator functions for NSFD schemes for differential equations satisfying a positivity condition.

22. A. Abrramowitz and I. A. Stegun, editors, *Handbook of Mathematical Functions* (U. S. National Bureau of Standards, Washington, DC, 1964).

23. G. A. Korn and T. M. Kern, *Mathematical Handbook for Scientific and Engineers* (McGraw-Hill, New York, 1961).

24. R. B. Potts, *Bulletin of the Australian Mathematical Society,* **35**, 43–48 (1987). Weierstrass elliptic difference equations.

25. R. E. Mickens, *Journal of Difference Equations and Applications* **24**, 1016–1022 (2018). (doi:10.1080/10236198.2018.1430792). A note on exact finite difference schemes for modified Lotka-Volterra differential equations.

Chapter 11

Applications and Related Topics

11.1 Introduction

The main purpose of this Chapter is to present a number of differential equations which arise from the modelling of phenomena in the natural universe and show how (valid) NSFD discretizations may be constructed. Again, it should be clearly understood and appreciated that the NSFD methodology does not allow the unique construction of a finite difference scheme for a given differential equation. Thoughtfulness, skill, experience, insight, and "luck" are needed to succeed. This is the reason why we call our work the "NSFD methodology" rather than a "NSFD theory." In fact, there may not exist a "NSFD theory."

Almost all of the results, discussions, and explanations to follow in this Chapter are based on previous published works of either the author or others. When appropriate, we provide the relevant references at the end of each section.

11.2 Stellar Structures

A critical issue in astrophysics is the determination of the internal structure and evolution of stars [1, 2]. We consider the basic equations for a spherically symmetric star where the following variables are identified as being relevant [3]:

$P(r)$: pressure

$M(r)$: mass

$L(r)$: luminosity

$\rho(r)$: density.

Note that for this case, all variables are functions of the radial distance, r.

This density $\rho(r)$ is given by the equation of state for an ideal gas [1]

$$\rho = \frac{P}{KT}, \tag{11.2.1}$$

where K is a known, positive constant. Denote by ϵ the non-negative "energy production function," i.e.,

$$\epsilon(\rho, T) \geq 0, \tag{11.2.2}$$

which depends on the density and temperature. With these designations, the following four, coupled, first-order differential equations allow us to calculate the pressure, mass, luminosity, and temperature as a function of the radial distance, r:

$$\frac{dP}{dr} = -\left(\frac{GM\rho}{r^2}\right) \tag{11.2.3}$$

$$\frac{dM}{dr} = 4\pi^2 \rho, \tag{11.2.4}$$

$$\frac{dL}{dr} = 4\pi r^2 \rho \epsilon, \tag{11.2.5}$$

$$\frac{dT}{dr} = -\left(\frac{\gamma - 1}{\gamma}\right)\left(\frac{GMT}{r^2}\right), \tag{11.2.6}$$

where G is the Newton gravitational constant [1] and γ is a constant [1] greater than one.

The physics of this system allows us to conclude that all four dependent variables must be non-negative functions of r. Further, the pressure $P(r)$ and the temperature $T(r)$ must each decrease monotonic from the center of the star to its surface, while the $M(r)$ and luminosity $L(r)$ both increase monotonic over the same interval in r.

A difficulty often arises when Eqs. (11.2.3)–(11.2.6) are numerically integrated [3]. Particular numerical schemes may give numerical solutions for which positivity and/or monotonicity are not satisfied. We now demonstrate that the NSFD methodology allows for the construction of a scheme such that the conditions of positivity and monotonicity hold [4].

In the calculations to follow, the four dependent variables will be calculated in the order

$$P \to M \to L \to T. \tag{11.2.7}$$

The following notation is used for the discretized variables

$$r \to r_m = (\Delta r)m, \quad \Delta r = h, \quad m = \text{integer}, \tag{11.2.8a}$$

$$[P(r), M(r), L(r), T(r)] \to [P_m, M_m, L_m, T_m], \tag{11.2.8b}$$

where P_m is the approximation to $P(r_m)$, etc.

First, consider Eq. (11.2.3) and replace the density by the equation of state for an ideal gas, as given in Eq. (11.2.1); doing this gives

$$\frac{dP}{dr} = -\left(\frac{G}{K}\right)\left(\frac{MP}{r^2 T}\right)$$

$$= -2\left(\frac{G}{K}\right)\left(\frac{MP}{r^2 T}\right) + \left(\frac{G}{K}\right)\left(\frac{MP}{r^2 T}\right). \tag{11.2.9}$$

Now use the NSFD discretization

$$\frac{P_{m+1} - P_m}{h} = -2\left(\frac{G}{K}\right)\left(\frac{M_m P_{m+1}}{r_m^2 T_m}\right) + \left(\frac{G}{K}\right)\left(\frac{M_m P_m}{r_m^2 T_m}\right), \tag{11.2.10}$$

to achieve positivity, i.e., solving for P_{m+1}, we find

$$P_{m+1} = \left[\frac{r_m^2 T_m + \left(\frac{hG}{K}\right) M_m}{r_m^2 T_m + 2\left(\frac{hG}{K}\right) M_m}\right] P_m. \tag{11.2.11}$$

Note that if

$$T_m > 0, \quad M_m > 0, \quad P_m > 0, \tag{11.2.12}$$

then

$$0 < P_{m+1} < P_m, \tag{11.2.13}$$

and both positivity and monotonicity holds.

Next, consider Eq. (11.2.4) and use the result of Eq. (11.2.1) to obtain

$$\frac{dM}{dr} = \left(\frac{4\pi}{K}\right)\left(\frac{r^2 P}{T}\right). \tag{11.2.14}$$

Now use the following discretization

$$\frac{M_{m+1} - M_m}{h} = \left(\frac{4\pi}{K}\right)\left(\frac{r^2 P_{m+1}}{T_m}\right), \tag{11.2.15}$$

or

$$M_{m+1} = M_m + \left(\frac{4\pi h}{K}\right)\left(\frac{r_m^2 P_{m+1}}{T_m}\right). \tag{11.2.16}$$

Note, P_{m+1} is known and is non-negative. This means that

$$M_{m+1} > M_m > 0, \tag{11.2.17}$$

and, as a consequence, M_m satisfies the required conditions of positivity and monotonicity.

The energy production function $\epsilon(\rho, T)$ appears in the differential equation for $L(r)$, i.e., Eq. (11.2.5), where

$$\epsilon(\rho, T) = \epsilon\left(\frac{P}{KT}, T\right) \geq 0. \tag{11.2.18}$$

Now define a new function $E(\rho, T)$ as follows

$$E(\rho, T) \equiv \rho\epsilon(\rho, T)$$
$$= \left(\frac{P}{KT}\right)\epsilon\left(\frac{P}{KT}, T\right)$$
$$\equiv E_1(P, T) \geq 0. \tag{11.2.19}$$

With this definition, Eq. (11.2.5) becomes

$$\frac{dL}{dr} = 4\pi r^2 E_1(P, T), \tag{11.2.20}$$

and the selected finite difference discretization is

$$\frac{L_{m+1} - L_m}{h} = 4\pi r_m^2 E_1(P_{m+1}, T_m) \tag{11.2.21}$$

or

$$L_{m+1} = L_m + 4\pi h r_m^2 E_1(P_{m+1}, T_m). \tag{11.2.22}$$

Again, we find that the correct positivity and monotonicity conditions hold, i.e.,

$$0 < L_m < L_{m+1}. \tag{11.2.23}$$

Write Eq. (11.2.6) as

$$\frac{dT}{dr} = -2\beta\left(\frac{MT}{r^2}\right) + \beta\left(MTr^2\right), \tag{11.2.24}$$

where

$$\beta = \left(\frac{\gamma - 1}{\gamma}\right)G, \tag{11.2.25}$$

and use the NSFD discretization

$$\frac{T_{m+1} - T_m}{h} = -2\beta\left(\frac{M_{m+1}T_{m+1}}{r_m^2}\right) + \beta\left(\frac{M_{m+1}T_m}{r_m^2}\right). \tag{11.2.26}$$

Solving for T_{m+1} gives

$$T_{m+1} = \left[\frac{r_m^2 + (h\beta)M_{m+1}}{r_m^2 + 2(h\beta)M_{m+1}}\right]T_m, \tag{11.2.27}$$

and it follows that

$$0 < T_{m+1} < T_m. \tag{11.2.28}$$

In summary, we have constructed NSFD discretizations for the four ordinary differential equations modelling the internal properties of a spherical star. All of the numerical solutions have the correct positivity and monotonicity properties. Observe that the scheme is sequential, i.e., the four dependent variables are numerically computed in a definite order, i.e.,

1) From (r_m, M_m, P_m), the value P_{m+1} is calculated.
2) Next, M_{m+1} is determined from (P_{m+1}, T_m, M_m).
3) The luminosity, L_{m+1} is then calculated from (L_m, T_m, P_{m+1}).
4) Finally, T_{m+1} is calculated from (T_m, M_{m+1}).

The results of this section are based on ref. [5].

References

1. S. Chandrasekhar, *Stellar Structures* (Dover Publications, New York, 1939).
2. M. Schwarzchild, *Structure and Evolution of the Stars* (Princeton University Press; Princeton, NJ; 1958).
3. C. A. Rouse, *Astronomy and Astrophysics* **71**, 95–101 (1979). On the radial oscillations of the 1968 nonstandard solar model.
4. My interest in this problem arose out of discussions with Dr. Carl Rouse in early 2002 in San Diego, CA. He indicated that many of the standard numerical procedures produced, on occasion, negative solutions.
5. R. E. Mickens, *Journal of Sound and Vibration* **265**, 1116–1120 (2003). A non-standard finite difference scheme for the equations modelling stellar structure.

11.3   The $x - y - z$ Model

The $x - y - z$ model is defined by the following three, coupled, nonlinear, first-order differential equations [1]

$$\frac{dx}{dt} = \mu(1 - x) + xy - xz, \tag{11.3.1}$$

$$\frac{dy}{dt} = \mu(1 - y) + yz - xy, \tag{11.3.2}$$

$$\frac{dz}{dt} = \mu(1 - z) + xz - yz, \tag{11.3.3}$$

where μ is a positive parameter. Note that these equations are invariant under the transformation

$$x \to y \to z \to x. \tag{11.3.4}$$

Further, they satisfy a conservation law [2]

$$\frac{dM}{dt} = \mu(3 - M), \quad M = x + y + z. \tag{11.3.5}$$

This set of equations were specifically constructed to test whether the application of the NSFD methodology could be used to formulate a discretization such that the corresponding conservation law, Eq. (11.3.5) holds exactly for the discrete, independent variables (x_k, y_k, z_k) [1], where

$$\begin{cases} t \to t_k = hk, \quad h = \Delta t, \quad k(0, 1, 2, \ldots); \\ \{x(t), y(t), z(t)\} \to (x_k, y_k, z_k). \end{cases} \tag{11.3.6}$$

First, inspection of Eqs. (11.3.1) to (11.3.3) allows one to conclude that their solutions satisfy a positivity condition [3], i.e.,

$$[x(0) = x_0 \geq 0, y(0) = y_0 \geq 0, z(0) = z_0 \geq 0]$$
$$\Rightarrow [x(t) \geq 0, y(t) \geq 0, z(t) \geq 0]. \tag{11.3.7}$$

For the discretization variables, (x_k, y_k, z_k), this becomes

$$[x_k \geq 0, y_k \geq 0, z_k \geq 0] \Rightarrow [x_{k+1} \geq 0, y_{k+1} \geq 0, z_{k+1} \geq 0]. \tag{11.3.8}$$

Second, the exact finite difference scheme for the conservation law, stated in Eq. (11.3.5), is [2]

$$\frac{M_{k+1} - M_k}{\phi} = \mu(3 - M_{k+1}), \tag{11.3.9}$$

where

$$\phi = \phi(\mu) = \frac{e^{\mu h} - 1}{\mu}, \tag{11.3.10}$$

and

$$M_k = x_k + y_k + z_k. \tag{11.3.11}$$

The results of Eq. (11.3.9) imply that all three denominator functions for Eqs. (11.3.1) to (11.3.3) must be the same as that given in Eq. (11.3.10), i.e.,

$$\frac{dx}{dt} \to \frac{x_{k+1} - x_k}{\phi}, \quad \frac{dy}{dt} \to \frac{y_{k+1} - y_k}{\phi}, \quad \frac{dz}{dt} \to \frac{z_{k+1} - z_k}{\phi}. \tag{11.3.12}$$

The NSFD methodology specifies that to achieve a positivity preserving scheme, at least one of the variables in a nonlinear term, proceeded by a negative sign, must be evaluated at the advanced discrete time $(k + 1)$. Further, if the same term appears in more than one ODE, it must be discretized in exactly the same manner. For the problem at hand, note that xz appears in both Eqs. (11.3.1) to (11.3.3), xy in Eqs. (11.3.1) and (11.3.2), and yz in Eqs. (11.3.2) and (11.3.3).

Our NSFD discretizations for the nonlinear terms are taken to be the expressions

$$\begin{pmatrix} xy \\ yz \\ zx \end{pmatrix} \to \begin{pmatrix} x_k y_{k+1} \\ y_k z_{k+1} \\ z_k x_{k+1} \end{pmatrix}. \tag{11.3.13}$$

This leads to the NSFD scheme

$$\frac{x_{k+1} - x_k}{\phi} = \mu(1 - x_{k+1}) + x_k y_{k+1} - z_k x_{k+1}, \tag{11.3.14}$$

$$\frac{y_{k+1} - y_k}{\phi} = \mu(1 - y_{k+1}) + y_k z_{k+1} - x_k y_{k+1}, \tag{11.3.15}$$

$$\frac{z_{k+1} - z_k}{\phi} = \mu(1 - z_{k+1}) + z_k x_{k+1} - y_k z_{k+1}. \tag{11.3.16}$$

Since $(x_{k+1}, y_{k+1}, z_{k+1})$ appear linearly, in these equations, they may be rewritten to the following matrix form

$$AX = D, \tag{11.3.17}$$

where

$$X = \begin{pmatrix} x_{k+1} \\ y_{k+1} \\ z_{k+1} \end{pmatrix}, \quad D = \begin{pmatrix} d_1 \\ d_2 \\ d_3 \end{pmatrix}, \quad A = \begin{pmatrix} a_1 & b_1 & c_1 \\ a_2 & b_2 & c_2 \\ a_3 & b_3 & c_3 \end{pmatrix}, \tag{11.3.18}$$

and

$$d_1 = x_k + \mu\phi, \quad d_2 = y_k + \mu\phi, \quad d_3 = z_k + \mu\phi, \tag{11.3.19}$$

$$a_1 = 1 + \mu\phi + \phi z_k, \quad b_1 = \phi x_k, \quad c_1 = 0, \tag{11.3.20}$$

$$a_2 = 0, \quad b_2 = 1 + \mu\phi + \phi x_k, \quad c_2 = \phi y_k, \tag{11.3.21}$$

$$a_3 = \phi z_k, \quad b_3 = 0, \quad c_3 = 1 + \mu\phi + \phi y_k. \tag{11.3.22}$$

Solving for X, in Eq. (11.3.17)

$$X = A^{-1}D, \tag{11.3.23}$$

or

$$\begin{pmatrix} x_{k+1} \\ y_{k+1} \\ z_{k+1} \end{pmatrix} = \left(\frac{1}{\det A}\right) \begin{pmatrix} b_2 c_3 & b_1 c_3 & b_1 c_2 \\ a_3 c_2 & a_1 c_3 & a_1 c_2 \\ a_3 b_2 & a_3 b_1 & a_1 b_2 \end{pmatrix} \begin{pmatrix} d_1 \\ d_2 \\ d_3 \end{pmatrix}, \tag{11.3.24}$$

where

$$\det A = \beta^3 + \phi\beta^2(x_k + y_k + z_k) \\ + \beta\phi^2(x_k y_k + y_k z_k + x_k z_k), \tag{11.3.25}$$

and

$$\beta = 1 + \mu\phi. \tag{11.3.26}$$

Close inspection of Eqs. (11.3.24), (11.3.25) and (11.3.26) allows the following conclusions to be reached:

1) The positivity condition given in Eq. (11.3.8) is satisfied.
2) There exist functions, (F, G, H), such that

$$x_{k+1} = F(x_k, y_k, z_k, \mu), \tag{11.3.27}$$

$$y_{k+1} = G(x_k, y_k, z_k, \mu), \tag{11.3.28}$$

$$z_{k+1} = H(x_k, y_k, z_k, \mu). \tag{11.3.29}$$

These functions can be explicitly calculated using the results in Eqs. (11.3.19) to (11.3.26).
3) Inspection of Eqs. (11.3.14) to (11.3.16) shows them to be semi-explicity schemes, with $(x_{k+1}, y_{k+1}, z_{k+1})$ appearing linearly. However, the schemes in Eqs. (11.3.27) to (11.3.29) are explicit.
4) Our conclusion is that the $x - y - z$ model can be discretized to form a valid NSFD scheme.

References

1. R. E. Mickens, *Journal of Difference Equations and Applications* **16**, 1501–1504 (2010). A note on a non-standard finite difference scheme for the Reluga $x - y - z$ model.
2. R. E. Mickens, *Journal of Biological Dynamics* **1**, 437–436 (2007). Numerical integration of population models satisfying conservation laws: NSFD methods.
3. H. R. Thieme, *Mathematics in Population Biology* (Princeton University Press; Princeton, NJ; 2003). See Appendix A.

11.4 Mickens' Modified Newton's Law of Cooling

Let an object of temperature T be placed into a uniform environment at temperature T_e. If $T > T_e$, then the object eventually cools to T_e, while if $T < T_e$, the object warms to T_e. It is also an experimental fact that this cooling or warming takes place in a finite time.

The Newton's law of cooling states that the dynamics is given by the following first-order, linear differential equation [1]

$$\frac{dT(t)}{dt} = -\lambda[T(t) - T_e], \quad T(0) = T_0, \tag{11.4.1}$$

where λ is a non-negative parameter. Using the variable

$$x(t) = T(t) - T_3, \tag{11.4.2}$$

Newton's equation becomes

$$\frac{dx}{dt} = -\lambda x \quad x(0) = x_0 = T_0 - T_e, \tag{11.4.3}$$

which has the solution

$$x(t) = x_0 e^{-\lambda t} \tag{11.4.4}$$

or

$$T(t) = T_e + (T_0 - T_e)e^{-\lambda t}. \tag{11.4.5}$$

Note that starting with $T(0) = T_0$, it takes an unlimited length of time to achieve the equilibrium value T_e, which is not consistent with the actual situation of finite-time dynamics. Thus, Newton's law should be modified.

However, before explicitly doing so, let us consider the general properties of such a law [2].

Using the variable x, we may express a general law for cooling/heating in the form

$$\frac{dx}{dt} = -\lambda f(x), \quad x(0) = x_0, \quad \lambda > 0. \tag{11.4.6}$$

To have $x = 0$ as the equilibrium value, then the function $f(x)$ must have the following properties:

1) $f(0) = 0$,
2) $f(x) > 0$, if $x > 0$,
3) $df(x)/dx > 0$, $x > 0$,
4) $f(-x) = -f(x)$.
5) $f(x)$ must be such that the dynamics is finite.

Observe that for the standard Newton's law, $f(x) = x$, and this function satisfies all of requirements except the fifth one.

A class of functions, $f(x)$, which has finite-time dynamics is

$$f(x) = [sgn(x)]x^p, \quad 0 < p < 1, \tag{11.4.7}$$

where

$$sgn(x) = \begin{cases} 1, & x > 0; \\ 0, & x = 0; \\ -1, & x < 0. \end{cases} \tag{11.4.8}$$

Also, the class of functions, $f(x)$,

$$\lambda f(x) = \lambda x + \epsilon [sgn(x)]x^p, \tag{11.4.9}$$

where $\lambda > 0$ and $\epsilon > 0$, has finite-time dynamics. (We demonstrate this later by calculating the explicit solutions for the special case where $p = \frac{1}{2}$.)

Note that the Mickens' modified Newton law of cooling/heating is the differential equation

$$\frac{dx}{dt} = -\lambda x - \epsilon [sgn(x)]x^p. \tag{11.4.10}$$

To proceed, take $p = \frac{1}{2}$, $x(0) = x_0 > 0$, i.e.,

$$\frac{dx}{dt} = -\lambda x - \epsilon x^{1/2}, \quad x(t_0) = x_0, \quad t_0 \geq 0. \tag{11.4.11}$$

With this selection

$$0 \leq x(t) \leq x_0, \quad t_0 < t < \infty. \tag{11.4.12}$$

Issue: Construct a NSFD scheme for Eq. (11.4.11), with the properties:

a)

$$x_k \geq 0 \Rightarrow x_{k+1} \geq 0. \qquad (11.4.13a)$$

b)

$$x_k > x_{k+1}. \qquad (11.4.13b)$$

c) Finite-time dynamics, i.e., there exists a $k = \bar{k}$, such that

$$x_{\bar{k}} > 0 \Rightarrow x_{\bar{k}+1} = 0. \qquad (11.4.13c)$$

One (valid) possibility is the scheme

$$\frac{x_{k+1} - x_k}{\phi_1(h, \lambda)} = -\lambda x_k - \epsilon \left(\frac{x_{k+1}}{\sqrt{x_k}} \right), \qquad (11.4.14)$$

where

$$\phi_1(h, \lambda) = \frac{1 - e^{-\lambda h}}{\lambda}. \qquad (11.4.15)$$

The denominator function, $\phi_1(h, \lambda)$, is determined from the linear term on the right-side of Eq. (11.4.11), and the nonlinear function, $\sqrt{x}$, is discretized using the nonlocal representation

$$\sqrt{x} \to \frac{x_{k+1}}{\sqrt{x_k}}. \qquad (11.4.16)$$

Solving Eq. (11.4.14) for x_{k+1} and simplifying the resulting expression gives

$$x_{k+1} = \left(\frac{\sqrt{x_k}}{\sqrt{x_k} + \epsilon \phi_1} \right) e^{-\lambda h} x_k. \qquad (11.4.17)$$

Inspection of Eq. (11.4.17) shows that it is consistent with requirements a) and b), but not c), i.e., the solution decays at least exponentially, but does not become zero for a large, but finite value of k.

It should be indicated that the NSFD scheme

$$\frac{x_{k+1} - x_k}{\phi_2(h, \lambda)} = -\lambda x_{k+1} - \epsilon \left(\frac{x_{k+1}}{\sqrt{x_k}} \right), \qquad (11.4.18)$$

where

$$\phi_2(h, \lambda) = \frac{e^{\lambda h} - 1}{\lambda}, \qquad (11.4.19)$$

also gives the result seen in Eq. (11.4.17).

It turns out that the ODE of Eq. (11.4.11) can be solved exactly and, as a consequence, an exact finite difference scheme can be determined.

To obtain the exact scheme, start by making the variable transformation

$$u = \sqrt{x}, \quad \text{or} \quad x = u^2 \Rightarrow \frac{dx}{dt} = 2u\frac{du}{dt}. \tag{11.4.20}$$

Substituting these results into Eq. (11.4.11) gives

$$\frac{du}{dt} = -\bar{\lambda}u - \bar{\epsilon}, \tag{11.4.21}$$

with

$$\bar{\lambda} = \frac{\lambda}{2}, \quad \bar{\epsilon} = \frac{\epsilon}{2}, \quad u(t_0) = u_0 = \sqrt{x(t_0)}. \tag{11.4.22}$$

The solution to Eq. (11.4.21), with the initial condition given in Eq. (11.4.22) is

$$u(t) = \left(u_0 + \frac{\bar{\epsilon}}{\bar{\lambda}}\right)e^{-\bar{\lambda}(t-t_0)} - \left(\frac{\bar{\epsilon}}{\bar{\lambda}}\right). \tag{11.4.23}$$

The corresponding exact finite difference scheme is obtained by making the following substitutions in the solution [3]

$$t_0 \to hk, \quad t \to h(k+1), \quad u_0 \to u_k, \quad u \to u_{k+1}. \tag{11.4.24}$$

Doing this gives

$$u_{k+1} = \left(u_k + \frac{\bar{\epsilon}}{\bar{\lambda}}\right)e^{\bar{\lambda}h} - \left(\frac{\bar{\epsilon}}{\bar{\lambda}}\right), \tag{11.4.25}$$

and this can be rearranged to give the expression

$$\frac{u_{k+1} - u_k}{\phi(h, \bar{\lambda})} = -\bar{\lambda}u_k - \bar{\epsilon}, \tag{11.4.26}$$

where

$$\phi(h, \bar{\lambda}) = \frac{1 - e^{-\bar{\lambda}h}}{\bar{\lambda}}. \tag{11.4.27}$$

Using the fact that $u_k = \sqrt{u_k}$, Eq. (11.4.26) becomes

$$\frac{\sqrt{x_{k+1}} - \sqrt{x_k}}{\phi} = -\bar{\lambda}\sqrt{x_k} - \bar{\epsilon}. \tag{11.4.28}$$

Now multiply the last equation $\left(\sqrt{x_{k+1}} + \sqrt{x_k}\right)$ and use, $\bar{\lambda} = \lambda/2$ and $\bar{\epsilon} = \epsilon/2$, to obtain

$$\frac{x_{k+1} - x_k}{\phi} = -\lambda\sqrt{x_k}\left(\frac{\sqrt{x_{k+1}} + x_k}{2}\right) - \epsilon\left(\frac{\sqrt{x_{k+1}} + x_k}{2}\right), \tag{11.4.29}$$

where ϕ is given in Eq. (11.4.27). This is the exact finite difference scheme for Eq. (11.4.11).

If we consider the exact difference scheme for u_k, Eq. (11.4.25), rather than that for $x_k = u_k^2$, Eq. (11.4.29), we see that this equation is a mapping from u_k to u_{k+1}. This mapping

$$u_{k+1} \equiv f(u_k) = u_k e^{-\bar{\lambda} h} - \left(\frac{\bar{\epsilon}}{\bar{\lambda}}\right)(1 - e^{1-\bar{\lambda} h}), \qquad (11.4.30)$$

is linear. It has a slope less than one and a negative u_{k+1} intercept. Therefore, all the mappings decrease, i.e.,

$$u_{k+1} < u_k, \qquad (11.4.31)$$

and when, for some $k = k_1$, u_{k_1} becomes negative, then the iteration stops, since the only mathematically relevant values of u_k are those for which $u_k \geq 0$. Thus, the exact scheme has finite discrete-time dynamics.

It should be noted that $u(t)$ should be written in the following correct mathematical form

$$u(t) = \begin{cases} \left(u_0 + \frac{\bar{\epsilon}}{\bar{\lambda}}\right) e^{-\bar{\lambda}(t-t_0)} - \left(\frac{\bar{\epsilon}}{\bar{\lambda}}\right), & 0 < t \leq t_c \\ 0, & t > t_c \end{cases} \qquad (11.4.32)$$

where

$$t_c = \left(\frac{1}{\bar{\lambda}}\right) \text{Ln}\left[\left(\frac{\bar{\lambda}}{\bar{\epsilon}}\right)\left(u_0 + \frac{\bar{\epsilon}}{\bar{\lambda}}\right) e^{\bar{\lambda} t_0}\right]. \qquad (11.4.33)$$

This explicitly illustrates the finite-time dynamics.

Finally, inspection of Eq. (11.4.29), suggests that the NSFD scheme for the original differential equation, Eq. (11.4.11), should take the following nonlocal discretizations for the x and $x^{1/2}$ terms

$$x \rightarrow \left(\frac{\sqrt{x_{k+1}} + \sqrt{x_k}}{2}\right)\sqrt{x_k}, \qquad (11.4.34a)$$

$$\sqrt{x} \rightarrow \frac{\sqrt{x_{k+1}} + \sqrt{x_k}}{2}. \qquad (11.4.34b)$$

References

1. P. A. Tipples and G. Muscat, *Physics for Scientists and Engineers*, Vol. 1 (Freeman, New York, 2008, 6th edition). See pp. 674–675.
2. K. Oyedeji and R. E. Mickens, *Bulletin of the American Physical Society* **63** (#19, 2018), Abstract: F01.00009, Modified law of cooling – continuum and discreet effects.
3. R. E. Mickens, *Nonstandard Finite Difference Models of Differential Equations* (World Scientific, Singapore, 1994). See Section 3.2.
4. R. E. Mickens, *Difference Equations: Theory and Applications* (Van Nostrand Reinhold, New York, 1990).

11.5   NSFD Schemes for $dx/dt = -\lambda x^\alpha$

The first-order, nonlinear, single-power ODE

$$\frac{dx}{dt} + -\lambda x^\alpha, \quad \lambda > 0, \quad x(t_0) = x_0 > 0, \tag{11.5.1}$$

has been extensively investigated; see, for example [1]. Since it is a separable equation its exact solution can be easily calculated. Assume $\alpha \neq 1$; then the solution is

$$x(t) = [x_0^{1-\alpha} - \lambda(1-\alpha)(t - t_0)]^{\frac{1}{1-\alpha}}. \tag{11.5.2}$$

Note for $\alpha = 1$, we have

$$\frac{dx}{dt} = -\lambda x, \quad x(t) = x_0 e^{-\lambda(t-t_0)}. \tag{11.5.3}$$

Further, for $0 < \alpha < 1$, $x(t)$ has finite-time dynamics, i.e., there exists a time t_c such that $x(t_0) > 0$ and for $0 < t \leq t_c$, $x(t)$ decreases monotonic to zero at t_c, $x(t_c) = 0$. For $t > t_c$, $x(t)$ is zero.

The corresponding exact finite difference scheme is ($\alpha \neq 1$)

$$x_{k+1} = [x_k^{1-\alpha} - \lambda(1-\alpha)h]^{\frac{1}{1-\alpha}}, \tag{11.5.4}$$

where $x_k = x(t_k)$, with $t_k = hk$, $h = \Delta t$. The expression, given by Eq. (11.5.4), can be converted into a discrete-time derivative formulation having the form ($\alpha \neq 1$)

$$\frac{x_{k+1} - x_k}{h} = F(x_k, x_{k+1}, \lambda, h), \tag{11.5.5}$$

where $F(\dots)$ is known, but a very complex function of its arguments [1].

In practice, what is generally needed is a valid NSFD scheme for the discretization of Eq. (11.5.1). A way to proceed is to consider the following scheme [1]

$$\frac{x_{k+1} - x_k}{h} = -\lambda x^{\alpha-1}[(1+\theta)x_{k+1} - \theta x_k], \tag{11.5.6}$$

where the parameter θ satisfies the condition

$$\theta \geq 0. \tag{11.5.7}$$

Solving for x_{k+1} gives

$$x_{k+1} = \left[\frac{1 + (h\lambda\theta)x_k^{\alpha-1}}{1 + h\lambda(1+\theta)x_k^{\alpha-1}} \right] x_k. \tag{11.5.8}$$

Inspection of this expression allows the following conclusions to be reached:

   i) $x_k > 0 \Rightarrow x_{k+1} > 0$.

ii) $x_{k+1} < x_k$.

iii) The fixed-point of the ODE, Eq. (11.5.1) and the NSFD discretization, Eq. (11.5.8) are both the same, namely, zero, i.e.,

$$\lim_{t\to\infty} x(t) = \lim_{k\to\infty} x_k = 0. \qquad (11.5.9)$$

Since the parameter θ is not *a priori* specified, then other considerations must be taken into account to determine its value. However, it is of great value to compare the exact scheme, Eq. (11.5.1), and the NSFD scheme, Eq. (11.5.8). This clearly shows the major difference in their mathematical structures.

Reference

1. L.-I. Wu Roeger and R. E. Mickens, *Journal of Difference Equations and Applications* (2011), DOI: 10.1080/ 10236198.2011.574622. Exact finite difference and nonstandard finite difference schemes for $dx/dt = -\lambda x^\alpha$.

11.6 Exact Scheme for Linear ODE's with Constant Coefficients

This topic could have been placed in the previous chapter. However, its methodology is a continuation of the work presented in Section 11.5 and thus this is the appropriate location for the issues it examines.

The results of this section follows clearly the work of Roeger and Mickens [1]. Also, see Potts [2].

Let D be the differential operator, defined as

$$Dx(t) \equiv \frac{dx(t)}{dt}. \qquad (11.6.1)$$

Therefore, the solution to

$$(D - a)x(t) = 0, \quad a = \text{constant}, \qquad (11.6.2)$$

is

$$x(t) = Ae^{at}. \qquad (11.6.3)$$

The corresponding exact finite difference scheme is

$$\frac{x_{k+1} - x_k}{\left(\frac{e^{ah}-1}{a}\right)} - ax_k = 0, \qquad (11.6.4)$$

where

$$t \to t_k = hk, \quad h = \Delta t, \quad x(t_k) = x_k. \tag{11.6.5}$$

If the difference operator $\Delta_\lambda(x_k)$ is defined as

$$\Delta_\lambda(x_k) = \frac{x_{k+1} - x_k}{\phi(h, \lambda)}, \tag{11.6.6}$$

where

$$\phi(h, \lambda) = \begin{cases} \frac{e^{\lambda h} - 1}{\lambda}, & \text{if } \lambda \neq 0, \\ h, & \text{if } \lambda = 0. \end{cases} \tag{11.6.7}$$

With this notation, Eq. (11.6.4) can be written as

$$(\Delta_a - a)x_k = 0. \tag{11.6.8}$$

Now consider a general n-th, linear homogeneous differential equations with constant coefficients. It can be expressed in the form

$$(D - \lambda_1)(D - \lambda_2) \cdots (D - \lambda_n)x(t) = 0, \tag{11.6.9}$$

where the $(\lambda_1, \lambda_2, \ldots, \lambda_n)$ are constants. This equation can also be written as

$$[D^n - (\lambda_1 + \lambda_2 + \cdots + \lambda_n)D^{n-1} + \cdots + (-)^n \lambda_1 \lambda_2 \ldots \lambda_n]x(t) = 0. \tag{11.6.10}$$

The following results were proved in Roeger and Mickens [1]:

(A) Assume that the λ's are unique and real, i.e.,

$$\lambda_i \neq \lambda_j, \quad \text{if } i \neq j; \quad \lambda_i = \lambda_i^*, \tag{11.6.11}$$

where $(*)$ is the complex conjugation operator. Then an exact finite difference scheme for Eq. (11.6.9) is

$$(\Delta_{\lambda_1} - \lambda_1)(\Delta_{\lambda_2} - \lambda_2) \cdots (\Delta_{\lambda_n} - \lambda_n)x_k = 0, \tag{11.6.12}$$

where the Δ_λ are the discrete operators defined in Eqs. (11.6.6) and (11.6.7).

(B) If now some of the λ's are complex conjugate, the following result holds:

Let $\lambda_1 = \alpha + i\beta$, with α and β real, with $\beta \neq 0$. Then an exact finite difference scheme for

$$(D - \lambda_1)(D - \lambda_1^*)x(t) = 0, \tag{11.6.13}$$

is

$$(\Delta_{\lambda_1} - \lambda_1)(\Delta_{\lambda_1^*} - \lambda_1^*)x_k = 0. \tag{11.6.14}$$

(C) Now consider the case for repeated roots. Assume that λ_1 has multiplicity m. The differential equation

$$(D - \lambda_1)^m x(t) = 0, \tag{11.6.15}$$

where λ_1 may be real or complex, and m is a positive integer, has the following exact finite difference scheme

$$(\Delta_{\lambda_1} - \lambda_1)^m x_k = 0. \tag{11.6.16}$$

Note that the general solution to Eq. (11.6.16) is [3]

$$x_k = (1 + \lambda_1 \phi)^k (a_0 + a_1 k + \cdots + a_{m-1} k^{m-1}), \tag{11.6.17}$$

where, from Eq. (11.6.7)

$$\phi = \phi(h, \lambda_1), \tag{11.6.18}$$

and $(a_0, a_1, \ldots, a_{m-1})$ are arbitrary constants.

Let us now examine the case for the non-homogeneous, linear equation, with constant coefficients, i.e.,

$$(D - \lambda_1)(D - \lambda_2) \cdots (D - \lambda_n) x(t) = f(t) \tag{11.6.19}$$

where $f(t)$ is a continuous function of t.

(D) For the non-homogeneous, linear differential equation, given by Eq. (11.6.19), let the general solution be written as

$$x(t) = x_H(t) + x_P(t), \tag{11.6.20}$$

where $x_H(t)$ is the general solution to the homogeneous equation, Eq. (11.6.9), and $x_P(t)$ is a particular solution to the non-homogeneous equation, Eq. (11.6.19). The exact finite difference scheme for Eq. (11.6.19) is

$$
\begin{aligned}
(\Delta_{\lambda_1} - \lambda_1)(\Delta_{\lambda_2} - \lambda_2) \cdots (\Delta_{\lambda_n} - \lambda_n) x_k \\
= (\Delta_{\lambda_1} - \lambda_1)(\Delta_{\lambda_2} - \lambda_2) \cdots (\Delta_{\lambda_n} - \lambda_n) x_P(t_k).
\end{aligned} \tag{11.6.21}
$$

To illustrate these results, consider the ODE

$$(D^2 - 3D + 2) x(t) = t \tag{11.6.22}$$

or

$$(D - 1)(D - 2) x(t) = t. \tag{11.6.23}$$

Its general solution is

$$x(t) = c_1 e^t + c_2 e^{2t} + \frac{t}{2} + \frac{3}{4}. \tag{11.6.24}$$

Based on Eq. (11.6.21), the exact finite difference representation for Eq. (11.6.22) is

$$(\Delta_1 - 1)(\Delta_2 - 2)x_k = t_k + \left[\frac{3}{2} - \frac{h(e^h + 2)}{(e^{2h} - 1)} \right]. \tag{11.6.25}$$

Observe that for small h, we have

$$\frac{3}{2} - \frac{h(e^h + 2)}{(e^{2h} - 1)} = O(h^2). \tag{11.6.26}$$

References

1. L.-I. Wu Roeger and R. E. Mickens, *Journal of Difference Equations and Applications* (2013), DOI: 10.1080/10236198.2013. 771635. Exact finite difference scheme for linear differential equation with constant coefficients.
2. R. B. Potts, *American Mathematical Monthly* **89**, 402–407 (1982). Differential and difference equations.
3. S. Elaydi, *An Introduction to Difference Equations*, 3rd edition (Springer, New York, 2005).

11.7   Discrete 1-Dim Hamiltonian Systems

Many physical systems can be isolated to an extent that they may be modeled as conservative systems, i.e., the total mechanical energy is constant [1]. A function which characterizes such systems is the Hamiltonian or energy function [1]. Our major purpose in this section is to provide an introduction to the discretization of Hamiltonian and from this derive the equations of motion. We discuss only 1-dim systems and base our presentation on Mickens [2], which follows the work of Greenspan [3, 4] and Mickens [5, 6]. See also the article by Washington [7].

To begin, consider a one-degree-of-freedom system for which the Newton force equation is [1]

$$\frac{d^2 x(t)}{dt^2} = f(x(t)). \tag{11.7.1}$$

With the definition

$$y = \frac{dx}{dt},$$
(11.7.2)

it follows that

$$\frac{d^2x}{dt^2} = y\frac{dy}{dx}.$$
(11.7.3)

Therefore

$$y\frac{dy}{dx} - f(x) = 0,$$
(11.7.4)

and integrating this expression gives

$$H(x,y) \equiv \frac{y^2}{2} + V(x) = \text{constant},$$
(11.7.5)

where $V(x)$ is the potential energy (PE) [1], i.e.,

$$PE \equiv V(x) = -\int^x f(z)dz.$$
(11.7.6)

The term $y^2/2$ is the kinetic energy (KE) [1], i.e.,

$$KE \equiv \frac{y^2}{2}.$$
(11.7.7)

The Hamiltonian for the system is the sum of the KE and PE, and is constant for a given set of initial conditions,

$$x(t_0)x_0, \quad y(t_0) = y_0,$$
(11.7.8)

which gives

$$H(x(t), y(t)) = H(x_0, y_0) = \frac{y_0^2}{2} + V(x_0).$$
(11.7.9)

Note that Eq. (11.7.1) can be written as the system of coupled differential equation

$$\frac{dx}{dt} = y, \quad \frac{dy}{dt} = f(x).$$
(11.7.10)

Also, it follows that

$$\frac{dx}{dt} = \frac{\partial H(x,y)}{\partial y}, \quad \frac{dy}{dt} = -\frac{\partial H(x,y)}{\partial x},$$
(11.7.11)

a result which holds for a general Hamiltonian function [1].

An interesting feature, for 1-dim Hamiltonian systems is that the equation of motion, Eq. (11.7.1) can be obtained by taking the time derivative of the Hamiltonian, i.e.,

$$H(x,y) = \frac{y^2}{2} + V(x) = \text{constant}$$
(11.7.12)

$$\frac{dH(x,y)}{dt} = y\frac{dy}{dt} + \frac{dV}{dx}\frac{dx}{dt}$$

$$= y\frac{d^2x}{dt^2} + (-f)y$$

$$= y\left[\frac{d^2x}{dt^2} - f\right]$$

$$= 0, \qquad\qquad (11.7.13)$$

and this gives Eq. (11.7.1).

The question now arises as to how to construct proper discretizations for the discrete-time formulation of Hamiltonian dynamics?

To begin, it is obvious that we have

$$\begin{cases} t \to t_k = hk, \quad h = \Delta t, \quad k = \text{integer}; \\ x(t) \to x_k, \quad y(t) \to y_k. \end{cases} \qquad (11.7.14)$$

But, let us look at $y(t) = dx(t)/dt$. This suggests that $y(t)$ should be replaced by something such as

$$y(t) = \frac{dx(t)}{dt} \to \frac{x_k - x_{k-1}}{\phi(h)}, \qquad (11.7.15a)$$

where

$$\phi(h) = h + O(h^2). \qquad (11.7.15b)$$

This implies that the discretization of $H(x,y)$ should only be a function of x_k and x_{k-1}, and not of x_k and y_k, i.e.,

$$H(x,y) \to \bar{H}(x_k, x_{k-1}). \qquad (11.7.16)$$

Now the issue is to determine $\bar{H}(x_k, x_{k-1})$. (From now on, we drop the overbar on $\bar{H}$.)

Since

$$H(x,y) = \text{constant} \Rightarrow H(x_k, x_{k-1}) = \text{constant}, \qquad (11.7.17)$$

it follows that

$$\Delta H(x_k, x_{k-1}) = 0, \qquad (11.7.18)$$

where for an arbitrary discrete function g_k, we define the operator Δ as

$$\Delta g_k = g_{k+1} - g_k. \qquad (11.7.19)$$

It follows from Eq. (11.7.18) that

$$H(x_{k+1}, x_k) = H(x_k, x_{k-1}). \qquad (11.7.20)$$

The discrete Hamiltonian also has the property [5, 6]

$$H(x_k, x_{k+1}) = H(x_{k-1}, x_k), \qquad (11.7.21)$$

that means that each term in the discretization of Hamiltonian must be such that overall it is invariant under the interchange of indices

$$k \leftrightarrow (k-1). \qquad (11.7.22)$$

Further, these results imply that the discrete Hamiltonian is, in general, a nonlinear, first-order difference equation, see Eq. (11.7.16), while the equation of motion, see Eq. (11.7.18), is a second-order difference equation. It also follows that the equation of motion will be invariant under the exchange of indices

$$(k+1) \leftrightarrow (k-1). \qquad (11.7.23)$$

This follows directly from Eqs. (11.7.20) and (11.7.21).

11.7.1 *Discrete Hamiltonian Construction*

A large variety of conservative physical phenomena can be modeled by fourth-order polynomial PE functions. We take this form to be expressed as

$$V(x) = V_0 + V_1 x + V_2 x^2 + V_3 x^3 + V_4 x^4, \qquad (11.7.24)$$

and, consequently, the Hamiltonian is

$$H(x, y) = \frac{y^2}{2} + V_0 + V_1 x + V_2 x^2 + V_3 x^3 + V_4 x^4, \qquad (11.7.25)$$

where V_0, V_1, V_2, V_3, V_4 are constants.

So, how do we select the proper (x_k, x_{k-1}) representations such that $H(x_k, x_{k-1}) = H(x_{k-1}, x_k)$?

First, consider y^2. Based on its definition as $y = dx/dt$, the obvious choice is

$$y_2 \to \left(\frac{x_k - x_{k-1}}{\phi} \right)^2. \qquad (11.7.26)$$

Observe that it is invariant under $k \leftrightarrow (k-1)$.

Second, the first and second terms in the PE can only be expressed as

$$V_0 \to V_0, \quad x \to \frac{x_k + x_{k-1}}{2}. \qquad (11.7.27)$$

Now, what do we construct for x^2? The most general expression is

$$x^2 \to \frac{x_k^2 + a x_k x_{k-1} + x_{k-1}^2}{2 + a}, \qquad (11.7.28)$$

where a is a positive constant. However, we will use the following minimalist representation [2, 6]

$$x^2 \to x_k x_{k-1}. \tag{11.7.29}$$

Likewise, for x^3 and x^4, we use the corresponding minimalist representations [2, 6]

$$x^3 \to \frac{x_k^2 x_{k-1} + x_k x_{k-1}^2}{2}, \tag{11.7.30}$$

$$x^4 \to x_k^2 x_{k-1}^2. \tag{11.7.31}$$

Putting all of this together, the following NSFD discretization is determined for the Hamiltonian function given in Eq. (11.7.25)

$$H(x_k, x_{k-1}) = \frac{1}{2}\left(\frac{x_k - x_{k-1}}{\phi}\right)^2 + V_0 + V_1\left(\frac{x_k + x_{k-1}}{2}\right)$$

$$+ V_2 x_k x_{k-1} + V_3\left(\frac{x_k^2 x_{k-1} + x_k x_{k-1}^2}{2}\right)$$

$$+ V_4 x_k^2 x_{k-1}^2. \tag{11.7.32}$$

Finally, as expected from the NSFD methodology, each term in $H(x_k, x_{k-1})$ is nonlocal, i.e., depends on both k and $(k-1)$, rather than just k.

11.7.2 *Discrete Equations of Motion for Eq.* (11.7.32)

Given $H(x_k, x_{k-1})$, the second-order equation of motion is calculated using

$$\Delta H(x_k, x_{k-1}) = 0. \tag{11.7.33}$$

The evaluation of the individual terms gives

$$\Delta V_0 = 0$$

$$\Delta\left(\frac{x_k + x_{k-1}}{2}\right) = \left(\frac{x_{k+1} - x_{k-1}}{2}\right) \cdot 1, \tag{11.7.34}$$

$$\Delta(x_k x_{k-1}) = (x_{k+1} - x_{k-1})x_k, \tag{11.7.35}$$

$$\left(\frac{1}{3}\right)\Delta\left(\frac{x_k^2 x_{k-1} + x_k x_{k-1}^2}{2}\right) = \left(\frac{x_{k+1} - x_{k-1}}{2}\right)\left(\frac{x_{k+1} + x_k + x_{k-1}}{3}\right)x_k \tag{11.7.36}$$

$$\left(\frac{1}{2}\right)\Delta(x_k^2 x_{k-1}^2) = (x_{k+1} - x_{k-1})\left(\frac{x_{k+1} + x_{k-1}}{2}\right)x_k^2. \tag{11.7.37}$$

Consequently, the second-order, discrete equation of motion corresponding to the Hamiltonian in Eq. (11.7.32) is

$$\frac{x_{k+1} - 2x_k + x_{k-1}}{\phi^2} + V_1 + 2V_2 x_k$$

$$+ 3V_3 \left(\frac{x_{k+1} + x_k + x_{k-1}}{3} \right) x_k$$

$$+ 4V_4 \left(\frac{x_{k+1} + x_{k-1}}{2} \right) x_k^2 = 0. \tag{11.7.38}$$

To obtain Eq. (11.7.38), we used

$$\Delta \left(\frac{x_k - x_{k-1}}{\phi} \right)^2 = (x_{k+1} - x_{k-1}) \left[\frac{x_{k+1} - 2x_k + x_{k-1}}{\phi^2} \right]. \tag{11.7.39}$$

Again, as expected, we see the expected $(k + 1) \leftrightarrow (k - 1)$ invariance of the individual terms in the equation of motion. This is a direct reflection of the $t \to -t$ invariance of Eq. (11.7.1).

11.7.3   *Non-Polynomial Potential Energy*

The rational PE function

$$V(x) = - \left(\frac{1}{1 + x^2} \right), \tag{11.7.40}$$

gives rise to the following equation of motion

$$\frac{d^2 x}{dt^2} = - \frac{2x}{(1 + x^2)^2}. \tag{11.7.41}$$

The corresponding discrete PE function is taken to be

$$V(x) \to V(x_k, x_{k-1}) = \frac{(-1)}{1 + x_k x_{k-1}}. \tag{11.7.42}$$

However, other possibilities exist, such as

$$-V(x_k, x_{k-1}) = \left(\frac{1}{2} \right) \left[\frac{1}{1 + x_k^2} + \frac{1}{1 + x_{k-1}^2} \right], \tag{11.7.43}$$

but for a variety of reasons, would not be used for the construction of the discrete-time Hamiltonian or equation of motion.

Applying the Δ operator to the Hamiltonian

$$H(x_k, x_{k-1}) = \left(\frac{1}{2} \right) \left(\frac{x_k - x_{k-1}}{\phi} \right)^2 + V(x_k, x_{k-1}), \tag{11.7.44}$$

gives, after some algebraic manipulation, the expression

$$\frac{x_{k+1} - 2x_k + x_{k-1}}{\phi^2} = - \left[\frac{2x_k}{(1 + x_{k+1}x_k)(1 + x_k x_{k-1})} \right]. \tag{11.7.45}$$

11.7.4   Two Interesting Results

In the process of discretizing the Hamiltonian functions and calculating the equation of motion for the "logarithmic" potential $V(x) = \text{Ln}\,|x|$, the following discrete form of the derivative of the logarithm was discovered

$$\frac{d}{dx}(\text{Ln}\,|x|) \to \frac{\text{Ln}\,|x_{k+1}| - \text{Ln}\,|x_{k-1}|}{x_{k+1} - x_{k-1}}. \qquad (11.7.46)$$

In a similar manner, the corresponding expression for a discrete representation of the $sgn(x)$ function,

$$sgn(x) = \begin{cases} 1, & \text{if } x > 0, \\ 0, & \text{if } x = 0, \\ -1, & \text{if } x < 0, \end{cases} \qquad (11.7.47)$$

is

$$sgn(x) \to \frac{|x_{k+1}| - |x_{k-1}|}{x_{k+1} - x_{k-1}}. \qquad (11.7.48)$$

The derivation of these two results are found, respectively, in Sections 9.5.3.2 and 9.5.3.3 of Mickens [2].

References

1. H. Goldsein, *Classical Mechanics* (Addison-Wesley; Reading, MA; 1980).

2. R. E. Mickens, *Difference Equations: Theory, Applications and Advanced Topics*, Third Edition (CRC Press; Roca Baton, FL; 2015). See Section 9.5.

3. D. Greenspan, *Discrete Models* (Addison-Wesley; Reading, MA; 1973).

4. D. Greenspan, *Arithmetic Applied Mathematics* (Pergamon, Oxford, 1980).

5. R. E. Mickens, *Journal of Sound and Vibration* **172**, 142–144 (1994). Construction of finite-difference scheme that exactly conserves energy for a mixed-parity oscillator.

6. R. E. Mickens, *Journal of Sound and Vibration* **240**, 587–591 (2001). A nonstandard finite difference scheme for conservative oscillators.

7. T. M. Washington, *Computer and Mathematics with Applications* **66**, 2251–2258 (2013). NSFD representations for polynomial terms appearing in the potential functions of 1-dimensional conservative systems.

11.8   Cube Root Oscillators

Nonlinear oscillators provide a wide range of nonlinear, second-order, differential equations to investigate mathematically [1, 2] and numerically. Two particular interesting examples are the *cube-root equation* (CE) and the inverse cube-root equation (ICRE). They are, respectively, [3, 4]

$$\text{CRE}: \frac{d^2x}{dt^2} + x^{1/3} = 0, \tag{11.8.1}$$

$$\text{ICRE}: \frac{d^2x}{dt^2} + \frac{1}{x^{1/3}} = 0. \tag{11.8.2}$$

The major task of this section is to construct and analyze NSFD discretizations of these two equations. Our presentation follows closely the published papers of Ehrhardt and Mickens [3] and Mickens and Wilkerson [4].

Both of these equations are explicit examples of so-called "truly nonlinear oscillator" [5]. These are oscillator differential equations, such as Eqs. (11.8.1) and (11.8.2), such that no linear term in x appears. A major consequence of this fact is that none of the standard perturbation procedures can be applied to calculate analytic approximate solutions expanded in some small parameter [5].

11.8.1   *Cube Root Oscillator*

The CRE can be written as the following pair of first-order coupled differential equations

$$\frac{dx}{dt} = y, \quad \frac{dy}{dt} = -x^{1/3}; \quad x(0) = A, \quad y(0) = 0. \tag{11.8.3}$$

In the $x - y$ phase-space, where $y = y(x)$, the trajectories are solutions to the first-order differential equation [1, 5]

$$\frac{dy}{dx} = -\frac{x^{1/3}}{y}. \tag{11.8.4}$$

Consequently, the first-integral, which is the energy or Hamiltonian of the system, is

$$H(x, y) = \frac{y^2}{2} + \left(\frac{3}{4}\right) x^{4/3} = \left(\frac{3}{4}\right) A^{4/3}. \tag{11.8.5}$$

For fixed A, Eq. (11.8.5) is a simple closed curve in the $x - y$ phase-space, and this result implies that all its solutions are periodic [1, 5].

Let T be the period, i.e.,

$$x(t + T) = x(t), \qquad (11.8.6)$$

then T can be calculated and has the value [3]

$$T(A) = (5.8696\ldots)A^{1/3}. \qquad (11.8.7)$$

Observe that the period is a function of the initial amplitude $x(0) = A$, which can be selected to be positive.

The paper of Ehrhardt and Mickens [3] constructs four finite difference schemes for the CRE. The first three discretizations are

$$(I): \quad \frac{x_{k+1} - x_k}{h} = y_k, \quad \frac{y_{k+1} - y_k}{h} = -x_k^{1/3}; \qquad (11.8.8)$$

$$(II): \quad \frac{x_{k+1} - x_k}{h} = y_k, \quad \frac{y_{k+1} - y_k}{h} = -x_{k+1}^{1/3}; \qquad (11.8.9)$$

$$(III): \quad \frac{x_{k+1} - x_k}{h} = y_{k+1}, \quad \frac{y_{k+1} - y_k}{h} = -x_k^{1/3}. \qquad (11.8.10)$$

Note that (I) is the standard forward-Euler scheme. This scheme generates numerical solutions that oscillate with increasing amplitude and thus (I) is not a valid numerical integration method. (See Figure 1, in ref. [3].) However, schemes (II) and (III) produced numerical solutions having constant amplitude, periodic solutions, as expected from theory. (See Figure 2, in ref. [3].)

A fourth, NSFD scheme was constructed based on the methodology presented in Section 11.7.

First, a discretization of $H(x, y)$, and Eq. (11.8.5), was done, and the following result was obtained

$$H(x_k, x_{k-1}) = \left(\frac{1}{2}\right)\left(\frac{x_k - x_{k-1}}{h}\right)^2 + \left(\frac{3}{4}\right)x_k^{2/3}x_{k-1}^{2/3}. \qquad (11.8.11)$$

The corresponding equation of motion was then calculated from the relation

$$\Delta H(x_k, x_{k-1}) = 0. \qquad (11.8.12)$$

After some algebraic manipulations to this equation, we finally obtain the expression

$$(IV): \quad \frac{x_{k+1} - 2x_k + x_{k-1}}{h^2} + \left[\frac{ex_k^{2/3}}{x_{k+1}^{2/3} + x_{k+1}^{1/3}x_{k-1}^{1/3} + x_{k-1}^{2/3}}\right] \cdot$$
$$\left(\frac{x_{k+1}^{1/3} + x_{k-1}^{1/3}}{2}\right) = 0. \qquad (11.8.13)$$

A detailed inspection of this result allows the following conclusions:

(i) The discretization is invariant under the interchange $(k+1) \leftrightarrow (k-1)$.

(ii) It is an implicit scheme.

(iii) The second-derivative is modeled by standard central-difference representations.

(iv) The $x^{1/3}$ term is modeled by an expression having a rather complex algebraic structure.

(v) Schemes (II) and (III) may be considered elementary NSFD discretizations, and for computational purposes may be preferred over scheme (IV), at least for the calculation of preliminary numerical results.

Note that schemes (I), (II), and (III), can be rewritten to the respective forms

$$(I): \qquad \frac{x_{k+1} - 2x_k + x_{k-1}}{h^2} + x_{k-1}^{1/3} = 0, \qquad (11.8.14)$$

$$(II), (III): \qquad \frac{x_{k+1} - 2x_k + x_{k-1}}{h^2} + x_k^{1/3} = 0. \qquad (11.8.15)$$

Examining these equations indicates why the standard forward-Euler scheme did not work; its equation of motion is not invariant under the exchange of indices, $(k + 1) \leftrightarrow (k - 1)$.

Finally, the authors of ref. [3] indicate that the use of the standard MATLAB one-step ODE solver **ode45** that is based on an explicit Runge-Kutta $(4, 5)$ formula, give numerical solutions where the oscillations damped with time.

11.8.2 *Inverse Cube-Root Oscillator*

The ICRE can be expressed in system form as [4]

$$\frac{dx}{dt} = y, \quad \frac{dy}{dt} = -\frac{1}{x^{1/3}}, \quad x(0) = A, \quad y(0) = 0. \qquad (11.8.16)$$

The trajectories, $y = y(x)$, in the $x - y$ phase-space are the solutions to the differential equation

$$\frac{dy}{dx} = -\frac{1}{yx^{1/3}}, \qquad (11.8.17)$$

and the associated Hamiltonian is

$$H(x, y) = \frac{y^2}{2} + \left(\frac{3}{2}\right) x^{2/3} = \left(\frac{3}{2}\right) A^{2/3}. \qquad (11.8.18)$$

Also, the period of the oscillator motion can be exactly calculated and its value is [4]

$$T(A) = \sqrt{3}\pi A^{2/3}. \qquad (11.8.19)$$

The NSFD scheme discrete form of the Hamiltonian, $H(x, y)$, will be taken as

$$H(x_k, x_{k-1}) = \left(\frac{1}{2}\right)\left(\frac{x_k - x_{k-1}}{h}\right)^2 + \left(\frac{3}{2}\right)x_k^{1/3}x_{k-1}^{1/3} = 0. \qquad (11.8.20)$$

The equation of motion is determined from the relation

$$\Delta H(x_k, x_{k-1}) = 0, \qquad (11.8.21)$$

and this gives

$$\frac{x_{k+1} - 2x_k + x_{k-1}}{h^2} + 3\left(\frac{x_{k+1}^{1/3} - x_{k-1}^{1/3}}{x_{k+1} - x_{k-1}}\right)x_k^{1/3} = 0. \qquad (11.8.22)$$

If we use the relation

$$a^3 - b^3 = (a - b)(a^2 + ab + b^2), \qquad (11.8.23)$$

where

$$a = x_{k+1}^{1/3}, \quad b = x_{k-1}^{1/3}, \qquad (11.8.24)$$

then Eq. (11.8.22) becomes

$$\frac{x_{k+1} - 2x_k + x_{k-1}}{h^2} + \left(\frac{3}{x_{k+1}^{2/3} + x_{k+1}^{1/3}x_{k-1}^{1/3} + x_{k-1}^{2/3}}\right)x_k^{1/3} = 0. \qquad (11.8.25)$$

Rewriting this expression as two, coupled, first-order, difference equations gives

$$\frac{x_{k+1} - x_k}{h} = y_k$$

$$\frac{y_k - y_{k-1}}{h} = -\left[\frac{3x_k^{1/3}}{x_{k+1}^{2/3} + x_{k+1}^{1/3}x_{k-1}^{1/3} + x_{k-1}^{2/3}}\right]. \qquad (11.8.26)$$

Finally, a standard scheme would give

$$\frac{x_{k+1} - x_k}{h} = y_k, \quad \frac{y_{k+1} - y_k}{h} = -\left(\frac{1}{x_k^{1/3}}\right), \qquad (11.8.27)$$

and

$$\frac{x_{k+1} - 2x_k + x_{k-1}}{h^2} + \frac{1}{x_{k-1}^{1/3}} = 0, \qquad (11.8.28)$$

and the last expression is not invariant under $(k + 1) \leftrightarrow (k - 1)$.

References

1. R. E. Mickens, *Oscillations in Planar Dynamic Systems* (World Scientific, Singapore, 1996).
2. R. E. Mickens, *Journal of Sound and Vibration* **111**, 515–518 (1986). A generalization of the method of harmonic balance.
3. M. Ehrhardt and R. E. Mickens, *Neural, Parallel and Scientific Computations* **16**, 179–188 (2008). Discrete models for the cube-root differential equation.
4. R. E. Mickens and D. Wilkerson, *Proceedings of Neural, Parallel, and Scientific Computations* **4**, 284–287 (2018). A nonstandard finite difference scheme for the inverse cube-root oscillator.
5. R. E. Mickens, *Truly Nonlinear Oscillations: Harmonic Balance, Parameter Expansions, Iteration, and Averaging Methods* (World Scientific, Singapore, 2010).

11.9 Alternative Methodologies for Constructing Discrete-Time Population Models

11.9.1 *Comments*

The construction of discretizations can take many forms, even within the context of finite difference techniques. This section explores two procedures for generating NSFD type schemes for differential equations. One uses an algebraic technique based on the approximation

$$1 - ax = \frac{1}{1 + ax} + O(x^2), \tag{11.9.1}$$

while the other use an exponential approximation method.

We apply these procedures to generate alternative discretizations of both single and multi-interacting populations [1], including the discrete-time Anderson-May model [2]. It should be indicated that the exponential approximation has been applied to both the heat equation [3] and the linear time-dependent Schrödinger [4] partial differential equations.

11.9.2 *Modified Anderson-May Models*

The original Anderson-May model can be written as [2]

$$S_{k+1} = B + S_k - \beta S_k I_k, \tag{11.9.2}$$

$$I_{k+1} = \beta S_k I_k, \tag{11.9.3}$$

where S_k and I_k represent the susceptible and infective populations at discrete-time k. These equations allow the disease dynamics to be determined as a function of k, given the initial-values, S_0 and I_0.

Examination of Eqs. (11.9.2) and (11.9.3) shows that they do not have a valid continuum limit as differential equations. We now modify these expressions to create difference formulations which do have a continuous limit.

Model A:

First, note that for small I_k, we have

$$S_k - \beta S_k I_k = S_k(1 - \beta I_k) \to S_k e^{-\beta I_k}, \tag{11.9.4}$$

$$\beta S_k I_k \to S_k(1 - e^{-\beta I_k}). \tag{11.9.5}$$

Making these substitutions in Eqs. (11.9.2) and (11.9.3) gives the modified Anderson-May equations

$$S_{k+1} = B + S_k e^{-\beta I_k}, \tag{11.9.6}$$

$$I_{k+1} = S_k(1 - e^{-\beta I_k}) + e^{-a} I_k, \tag{11.9.7}$$

$$R_{k+1} = (1 - e^{-a})I_k + R_k, \tag{11.9.8}$$

where transfer out of the infective population to a removed population, R_k, has been included. The parameter a is a positive, i.e., $s > 0$, and with this understanding the last three equations have the following interpretation:

(i) The susceptible population has a net birthrate, $B > 0$. Its value, S_{k+1}, at the $(k+1)$th time is equal to the new additions from birth, and the susceptibles, who were not infected, i.e., the term $S_k \exp(-\beta I_k)$. Thus, the exponential term, $\exp(-\beta I_k)$, plays the role of a probability for not becoming infective.

(ii) The population of infectives at time $(k+1)$ is equal to these susceptible members who became infected, added to the number of the infected population at time k who did not transfer out to the removed population, R_k. The exponential factor, $\exp(-a)$, is the probability for an infective person not to transfer to the removed class.

(iii) The result given in Eq. (11.9.8) is now obvious in its interpretation.

(iv) Adding Eqs. (11.9.6) to (11.9.8) gives

$$\Delta P_k = B, \quad P_k = S_k + I_k + R_k. \tag{11.9.9}$$

Thus, the total population increases by an amount equal to the birthrate.

Appropriate continuum limits can be formulated for Eqs. (11.9.6) to (11.9.8). First, assume that for $h = \Delta t$, we have

$$B \to h\bar{B}, \quad \beta \to h\bar{\beta}, \quad a \to h\bar{a}. \qquad (11.9.10)$$

Then for "small" h, we have for Eq. (11.9.6)

$$S_{k+1} = h\bar{B} + S_k[1 - h\bar{\beta}I_k + O(h^2)], \qquad (11.9.11)$$

which can be rewritten to the form

$$\frac{S_{k+1} - S_K}{h} = \bar{B} - \bar{\beta}S_k I_k + O(h). \qquad (11.9.12)$$

Therefore, taking the limit $(h \to 0)$ gives the differential equation

$$\frac{dS}{dt} = \bar{B} - \bar{\beta}SI. \qquad (11.9.13)$$

Carrying out the same steps for Eqs. (11.9.7) and (11.9.8) gives

$$\frac{dI}{dt} = \bar{\beta}SI - \bar{a}I, \qquad (11.9.14)$$

$$\frac{dR}{dt} = \bar{a}I, \qquad (11.9.15)$$

and adding these differential equations produces the result

$$\frac{dP}{dt} = \bar{B}, \quad P(t) = S(t) + I(t) + R(t). \qquad (11.9.16)$$

Model B:

A second modified Anderson-May model can be constructed using the relations

$$S_k - \beta S_k I_k = S_k(1 - \beta I_k) \to \frac{S_k}{1 + \beta I_k}, \qquad (11.9.17)$$

$$\beta S_k I_k \to \frac{\beta S_k I_k}{1 + \beta I_k}. \qquad (11.9.18)$$

With these replacements, the original Anderson-May equations become

$$S_{k+1} = B + \left(\frac{1}{1 + \beta I_k}\right) S_k, \qquad (11.9.19)$$

$$I_{k+1} = \left(\frac{\beta I_k}{1 + \beta I_k}\right) S_k + e^{-a} I_k, \qquad (11.9.20)$$

$$R_{k+1} = R_k + (1 - e^{-a}) I_k, \qquad (11.9.21)$$

with

$$\Delta P_k = B, \quad P_k = S_k + I_k + R_k. \tag{11.9.22}$$

The corresponding continuum limit, differential equations are the same as those in Eqs. (11.9.13) to (11.9.15).

In summary, we have shown how to modify the Anderson-May model such that differential equations limits exist. These new coupled, discrete-time equation may be considered NSFD schemes for the corresponding differential equations. An interesting feature is that both discrete-time modifications have the same continuum limits.

11.9.3   *Discrete Exponentialization*

The presentation given in this section is based on work done by Bhattachchargya [3] and Turchin [1, see pp. 55–66]. We have generalized their results [5].

Ricker Equation

Consider the logistic equation

$$\frac{dx}{dt} = rx\left(1 - \frac{x}{x^*}\right). \tag{11.9.23}$$

This equation is used to model the dynamics of a single self-interacting population [1]. The constants, r and x^*, are, respectively, the birthrate for small populations and the long-time equilibrium, steady-state population.

Dividing by x and using

$$\frac{dx}{x} = d\operatorname{Ln}(x), \tag{11.9.24}$$

gives

$$\frac{d}{dt}\operatorname{Ln}(x) = r\left(1 - \frac{x}{x^*}\right). \tag{11.9.25}$$

If $h = \Delta t$ is sufficiently small, then we may replace Eq. (11.9.25) by the approximation

$$\operatorname{Ln}\left(\frac{x_{k+1}}{x_k}\right) = hr\left(1 - \frac{x_k}{x^*}\right), \tag{11.9.26}$$

and it can be rewritten as

$$x_{k+1} = x_k \exp\left[hr\left(1 - \frac{x_k}{x^*}\right)\right], \tag{11.9.27}$$

which is the discrete-time Ricker equation for single population growth [1].

We can modify the Ricker equation by replacing the step-size, h, by a denominator function, $\phi(r, h)$, i.e., [7]

$$h \to \phi(r, h) = \frac{e^{rh} - 1}{r}. \tag{11.9.28}$$

Consequently, the discrete exponentialization of the logistic equation becomes the following NSFD modification of the Ricker equation

$$x_{k+1} = x_k \exp\left[\psi(rh)\left(1 - \frac{x_k}{x^*}\right)\right], \tag{11.9.29}$$

where

$$\psi(rh) = r\phi(r, h) = e^{rh} - 1. \tag{11.9.30}$$

Single Population Models

Single, self-interacting population models have the form

$$\frac{dx}{dt} = xf(x). \tag{11.9.31}$$

From

$$\frac{d\,\mathrm{Ln}(x)}{dt} = f(x), \tag{11.9.32}$$

using the arguments above, we obtain

$$x_{k+1} = x_k \exp[hf(x_k)], \tag{11.9.33}$$

which can be generalized to

$$x_{k+1} = x_k \exp[\phi(h, f_0)f(x_k)], \tag{11.9.34}$$

where

$$f_0 = f(0), \tag{11.9.35}$$

and

$$\phi(h, f_0) = \begin{cases} \frac{e^{f(0)h} - 1}{f(0)}, & \text{if } f(0) \neq 0, \\ h, & \text{if } f(0) = 0. \end{cases} \tag{11.9.36}$$

For special cases, such as the logistic equation, other discretizations are possible. An example is the following scheme for the logistic equation

$$x_{k+1} = x_k \exp\left[r\phi\left(1 - \frac{x_{k+1}}{x^*}\right)\right]. \tag{11.9.37}$$

This difference equation can be "exactly" solved in terms of the Lambert-W function [6].

Two-Interacting Populations

The modelling equations for this case takes the form [1]

$$\frac{dx}{dt} = xf(x,y), \quad \frac{dy}{dt} = yg(x,y). \tag{11.9.38}$$

The direct, discrete exponentialization scheme is

$$x_{k+1} = x_k \exp[\phi_1 f(x_k, y_k)], \tag{11.9.39}$$

$$y_{k+1} = y_k \exp[\phi_2 g(x_k, y_k)], \tag{11.9.40}$$

where

$$a_1 = f(0,0), \quad a_2 = g(0,0), \tag{11.9.41}$$

and

$$\phi_1 = \frac{e^{a_1 h} - 1}{a_1}, \quad \phi_2 = \frac{e^{a_2 h} - 1}{a_2}. \tag{11.9.42}$$

An example of such a system is the predator-prey model [1]

$$\frac{dx}{dt} = x(a_1 - b_1 x - c_1 y), \tag{11.9.43}$$

$$\frac{dy}{dt} = y(-a_2 + c_2 x), \tag{11.9.44}$$

for which

$$f(x,y) = a_1 - b_1 x - c_1 y, \quad g(x,y) = -a_2 + c_2 x. \tag{11.9.45}$$

A possible NSFD exponential discretization is

$$x_{k+1} = x_k \exp[\phi_1(a_1, h)(a_1 - b_1 x_{k+1} - c y_k], \tag{11.9.46}$$

$$y_{k+1} = y_k \exp[\phi_2(a_2, h)(-a_2 + c_2 x_{k+1}]. \tag{11.9.47}$$

Note that this scheme always obeys the positivity condition, i.e.,

$$(x_k \geq 0, y_k \geq 0) \Rightarrow (x_{k+1} \geq 0, y_{k+1} \geq 0). \tag{11.9.48}$$

For N-interacting populations where

$$x(t) = \begin{pmatrix} x_1(t) \\ x_2(t) \\ \vdots \\ x_n(t) \end{pmatrix}; \frac{dx_i(t)}{dt} = x_i f_i(x), \quad i = (1, \ldots, n), \tag{11.9.49}$$

then the above construction can be extended to this system of equations to obtain

$$x_i(k+1) = x_i(k) \exp[\phi_i(h) f_i(x)], \tag{11.9.50}$$

where

$$\phi_i(h) = \frac{e^{f_i(0)h} - 1}{f_i(0)} \tag{11.9.51}$$

for $i = (1, 2, 3, \ldots, n)$.

References

1. P. Turchin, *Complex Population Dynamics* (Princeton University Press; Princeton, NJ; 2003).
2. R. Anderson and R. May, *New Scientist* (November 18, 1982), 410–415. The logic of vaccination.
3. M. C. Bhattacharya, *Applied Mathematical Modelling* **10**, 68–70 (1986). A new improved finite difference equation for heat transfer during transient change.
4. R. E. Mickens, *Journal of Difference Equations and Applications* **12**, 313–320 (2006). A nonlinear NSFD scheme for the linear, time-dependent Schrödinger equation.
5. R. E. Mickens, *Abstracts of Papers Presented at the American Mathematical Society* **39** (Number 1, Issue 191, Winter 2018). Abstract 1135-gz-539, p. 334. Some methodologies for constructing discrete-time population models.
6. R. M. Corless, et al., *Advances in Computational Mathematics* **5**, 239–265 (1996). On the Lambert-W function.
7. R. E. Mickens, *Nonstandard Finite Difference Models of Differential Equations* (World Scientific, Singapore, 1994).

11.10   Interacting Populations with Conservation Laws

11.10.0   *Comments*

Interacting population models are ubiquitous in the natural and engineering sciences. They are used to analyze, understand, and predict the evolutions of the systems they supposedly represent. Populations can range from actual numbers of discrete individuals, such as atoms, to continuous based concepts, such as densities. A critical feature in both cases is that they are generally non-negative quantities. This section presents a number of interacting population models with the additional requirement that the system satisfies a conservation law.

Conservation laws are well known in the physical sciences [1] and also appears in the bio-sciences [2, 3, 4]. The main purpose of this section is to show how NSFD schemes can be constructed for the coupled ODE's modelling important phenomena in the bio-sciences where the dependent variables also satisfy a conservation law [2, 3, 4].

The next section explains how we define a conservation law [4]. This is followed by examining the NSFD discretization of some interacting population models taken primarily from the bio-sciences.

11.10.1 *Conservation Laws*

Consider a system of N-interacting populations, denoted by

$$P(t)^T = (P_1(t), P_2(t), \ldots, P_n(t)), \tag{11.10.1}$$

and let $F(P)$ represent their interaction terms, where

$$F(P)^T = (f_1(P), f_2(P), \ldots, f_n(P)), \tag{11.10.2}$$

such that the system dynamics is given by

$$\frac{dP}{dt} = F(P). \tag{11.10.3}$$

Further, it is assumed the $F(P)$ has the mathematical properties such that a "positivity condition" holds, i.e.,

$$P_i(0) \geq 0 \Rightarrow P_i(t) \geq 0 \quad \text{for } t > 0; \quad i = 1, 2, \ldots, N. \tag{11.10.4}$$

Let $\bar{P}(t)$ be the total population defined as

$$\bar{P}(t) = P_1(t) + P_2(t) + \cdots + P_n(t). \tag{11.10.5}$$

For our purposes, there are three classes of conservation laws and we label them, respectively, as [4]

- direct conservation laws
- generalized conservation laws
- sub-conservation laws.

Direct conservation law

If the total population, $\bar{P}(t)$, satisfies

$$\frac{d\bar{P}(t)}{dt} = 0 \Rightarrow \bar{P}(t) = \text{constant}, \tag{11.10.6}$$

then the system of interacting populations is said to satisfy a *direct conservation law*. Note that the total population is constant and that constant value is the sum of the initial values of all the individual populations, i.e.,

$$\begin{aligned}
\bar{P}(t) &= P_1(t) + P_2(t) + \cdots + P_n(t) \\
&= P_1(0) + P_2(0) + \cdots + P_n(0).
\end{aligned} \tag{11.10.7}$$

The positivity condition implies that if one or more populations decrease, then some of the other populations must increase.

Generalized conservation law

Suppose that the total population satisfies the following differential equation

$$\frac{d\bar{P}(t)}{dt} = a_1 - b_1 \bar{P}(t), \tag{11.10.8}$$

where a_1 and b_1 are positive constants. In this case, we said that the system of interacting populations satisfies a *generalized conservation law*.

The solution to Eq. (11.10.8) is

$$\bar{P}(t) = \left(\frac{a_1}{b_1}\right) + \left[\bar{P}(0) - \left(\frac{a_1}{b_1}\right)\right] e^P - b_1 t. \tag{11.10.9}$$

Note that

$$\bar{P}(\infty) = \underset{t\to\infty}{\text{Lim}}\, \bar{P}(t) = \frac{a_1}{b_1}. \tag{11.10.10}$$

Sub-conservation law

Assume $\bar{P}(t) > 0$, and not equal to a constant. Further, assume that a sub-set of the population, $P_S(t)$, where

$$P_S(t) = P_{1S}(t) + P_{2S}(t) + \cdots + P_{mS}(t), \tag{11.10.11a}$$

$$2 \le m < N \tag{11.10.11b}$$

has the feature that

$$\frac{dP_S(t)}{dt} = 0 \Rightarrow P_S(t) = \text{constant}, \tag{11.10.12}$$

or

$$\frac{dP_S(t)}{dt} = a_2 - b_2 P_S(t), \tag{11.10.13}$$

which implies

$$P_S(t) = \left(\frac{a_2}{b_2}\right) + \left[P_S(0) - \left(\frac{a_2}{b_2}\right)\right] e^{-b_2 t}. \tag{11.10.14}$$

If these conditions hold, then the system is said to have a *sub-conservation law*.

The remainder of this section will show that many of the familiar interacting populations satisfy a conservation law. We will use this fact, along with the positivity, to construct valid NSFD schemes for the associated system differential equations. Two other NSFD related concepts will also play important roles in these discretizations, namely, dynamics consistency [5] and how to construct denominator functions [6].

11.10.2   Chemostate Model

A simple one-resource, one-species chemostat model is described by two coupled, nonlinear, first-order differential equations

$$\frac{ds}{dt} = \mu - \mu s - rns, \qquad (11.10.15)$$

$$\frac{dn}{dt} = -\mu n + rns, \qquad (11.10.16)$$

where μ and r are positive parameters, s is the resource population, and n is the consumer of the resource. Adding the two equations gives

$$\frac{dP}{dt} = \mu(1 - P), \quad P(t) = s(t) + n(t). \qquad (11.10.17)$$

The exact discretization for Eq. (11.10.17) is

$$\frac{P_{k+1} - P_k}{\phi} = \mu(1 - P_{k+1}), \qquad (11.10.18)$$

with

$$\phi(h, \mu) = \frac{e^{\mu h} - 1}{\mu}. \qquad (11.10.19)$$

The results from the last two equations imply that the denominator functions for the discretizations of Eqs. (11.10.15) and (11.10.16) must be the same and equal to $\phi(h, \mu)$ given in Eq. (11.10.19). Consequently, the NSFD schemes are

$$\frac{s_{k+1} - s_k}{\phi} = \mu - \mu s_{k+1} - rn_k s_{k+1}, \qquad (11.10.20)$$

$$\frac{n_{k+1} - n_k}{\phi} = -\mu n_{k+1} + rn_k s_{k+1}. \qquad (11.10.21)$$

Note that the $(-\mu s)$ and $(-rns)$ terms are modeled, respectively, so that the conservation law is satisfied and that positivity holds. The (rns) term in the ODE for $n(t)$ must be exactly equal to the same term in the ODE for the $s(t)$ differential equation; likewise, the $(-\mu n)$ term is constructed to satisfy positivity.

Solving for s_{k+1} and n_{k+1} gives

$$s_{k+1} = \frac{s_k + \mu\phi}{1 + \mu\phi + r\phi n_k}, \qquad (11.10.22)$$

$$n_{k+1} = \left(\frac{1 + r\phi s_{k-1}}{1 + \mu\phi}\right) n_k. \qquad (11.10.23)$$

Thus, we have constructed a NSFD scheme for the simple chemostat model such that both positivity and the conservation law holds.

It is important to understand that the scheme *satisfied exactly* the conservation law. However, the nonstandard discretizations for $s(t)$ and $n(t)$ are not exact schemes.

11.10.3 *SIR Model*

One of the first and best studied models for the spread of disease is the SIR model

$$\frac{dS}{dt} = -\beta SI, \quad \frac{dI}{dt} = \beta SI - \gamma I, \quad \frac{dR}{dt} = \gamma I, \tag{11.10.24}$$

(S, I, R) are, respectively, the susceptible, inflected, and recovered populations, and (β, γ) are positive parameters.

Adding these three equations gives

$$\frac{dP(t)}{dt} = 0 \Rightarrow P(t) = S(t) + I(t) + R(t) = \text{constant}, \tag{11.10.25}$$

where the initial conditions are generally taken to be

$$S(0) = S_0 > 0, \quad I(0) = I_0 > 0, \quad R(0) = R_0 = 0, \tag{11.10.26a}$$

$$S_0 \gg I_0. \tag{11.10.26b}$$

The following argument can be used to determine the denominator function. If $S(t) = 0$, the $I(t)$ satisfies the equation

$$\frac{dI}{dt} = -\gamma I, \quad I(0) = I_0, \tag{11.10.27}$$

and this ODE has the exact discretization

$$\frac{I_{k+1} - I_k}{\phi} = -\gamma I_k, \quad \phi(h, \gamma) = \frac{1 - e^{-\gamma h}}{\gamma}. \tag{11.10.28}$$

Therefore, a positivity preserving scheme for Eqs. (11.10.24) is

$$\frac{S_{k+1} - S_k}{\phi} = -\beta S_{k+1} I_k, \tag{11.10.29}$$

$$\frac{I_{k+1} - I_k}{\phi} = \beta S_{k+1} I_k - \gamma I_k, \tag{11.10.30}$$

$$\frac{R_{k+1} - R_k}{\phi} = \gamma I_k, \tag{11.10.31}$$

along with the conservation law

$$\frac{P_{k+1} - P_k}{\phi} = 0. \tag{11.10.32}$$

Again, observe that if the three ODE discretizations did not have the same denominator functions, then the discrete version of the conservation law would not hold.

Solving for $(S_{k+1}, I_{k+1}, R_{k+1})$, using the expressions in Eqs. (11.10.29) to (11.10.31), gives

$$S_{k+1} = \frac{S_k}{1 + (\beta\phi)I_k}, \tag{11.10.33}$$

$$I_{k+1} = (e^{-\gamma h})I_k + (\beta\phi)S_{k+1}I_k, \tag{11.10.34}$$

$$R_{k+1} = R_k + (\gamma\phi)I_k. \tag{11.10.35}$$

Notice that to carry out the calculation of $(S_{k+1}, I_{k+1}, R_{k+1})$ we must

- first, calculate S_{k+1} from S_k and I_k;
- second, determine I_{k+1} from S_{k+1} and I_k;
- third, evaluate R_{k+1} from I_k and R_k.

11.10.4 SEIR Model with Net Birthrate

The four dependent variables for this system are (i) $S(t)$, the susceptible population, (ii) $E(t)$, the infected population, (iii) $I(t)$, the infectious population, and (iv) $R(t)$, the removed population.

This model is represented by four coupled, first-order, ODE's. For our purposes, we use the following set

$$\frac{dS}{dt} = B - mS - \beta\left(\frac{I}{N}\right)S, \tag{11.10.36}$$

$$\frac{dE}{dt} = \beta\left(\frac{I}{N}\right)S - gE - mE, \tag{11.10.37}$$

$$\frac{dI}{dt} = gE - mI - \gamma I, \tag{11.10.38}$$

$$\frac{dR}{dt} = \gamma I - mR, \tag{11.10.39}$$

where the parameters (B, g, m, β, γ) are non-negative, and the total population $N(t)$ is

$$N(t) = S(t) + E(t) + I(t) + R(t), \tag{11.10.40}$$

which satisfies the conservation law

$$\frac{dN}{dt} = B - mN(t). \tag{11.10.41}$$

An exact discretization of Eq. (11.10.41) is

$$\frac{N_{k+1} - N_k}{\phi} = B - mN_k, \quad \phi = \frac{1 - e^{-mh}}{m}. \tag{11.10.42}$$

Further, positivity satisfying NSFD schemes for Eqs. (11.10.36) and (11.10.39) give the expressions

$$\frac{S_{k+1} - S_k}{\phi} = B - mS_k - \beta\left(\frac{I_k}{N_k}\right)S_{k+1}, \tag{11.10.43}$$

$$\frac{E_{k+1} - E_k}{\phi} = \beta\left(\frac{I_k}{N_k}\right)S_{k+1} - gE_{k+1} - mE_k, \tag{11.10.44}$$

$$\frac{I_{k+1} - I_k}{\phi} = gE_{k+1} - mI_k - \gamma I_{k+1}, \tag{11.10.45}$$

$$\frac{R_{k+1} - R_k}{\phi} = \gamma I_{k+1} - mR_k. \tag{11.10.46}$$

If we now solve for the variables evaluated at the $(k+1)$ discrete-time level, we obtain

$$S_{k+1} = \frac{B\phi + (e^{-mh})S_k}{1 + (\beta\phi)\left(\frac{I_k}{N_k}\right)}, \tag{11.10.47}$$

$$E_{k+1} = \frac{(e^{-mh})E_k + (\beta\phi)S_{k+1}\left(\frac{I_k}{N_k}\right)}{1 + g\phi}, \tag{11.10.48}$$

$$I_{k+1} = \frac{(e^{-mh})I_k + (g\phi)E_{k+1}}{1 + \gamma\phi}, \tag{11.10.49}$$

$$R_{k+1} = (\gamma\phi)I_{k+1} + (e^{-mh})R_k. \tag{11.10.50}$$

We can calculate $(S_{k+1}, E_{k+1}, I_{k+1}, R_{k+1})$ by using the following procedure:

(1) calculate S_{k+1} from knowledge of (S_k, E_k, I_k, R_k);
(2) calculate E_{k+1} from $(S_{k+1}, S_k, E_k, I_k, R_k)$;
(3) calculate I_{k+1} from (I_k, E_{k+1});
(4) calculate R_{k+1} from (I_{k+1}, R_k).

11.10.5 *Criss-Cross Model* [7]

A model of venereal disease infection, with permanent immunity for the removed class, is an example of a so-called criss-cross model. The relevant equations are

$$\frac{dS}{dt} = -r_1 S\bar{I}, \quad \frac{dI}{dt} = r_1 S\bar{I} - a_1 I, \quad \frac{dR}{dt} = a_1 I, \tag{11.10.51}$$

$$\frac{d\bar{S}}{dt} = -r_2\bar{S}I, \quad \frac{d\bar{I}}{dt} = r_2\bar{S}I - a_2\bar{I}, \quad \frac{d\bar{R}}{dt} = a_2\bar{I}, \tag{11.10.52}$$

where (r_1, a_1, S, I, R) and $(r_2, a_2, \bar{S}, \bar{I}, \bar{R})$ are the parameters and variables corresponding to the female and male populations.

For this model, there are two conservation laws, i.e.,

$$N_1(t) = S(t) + I(t) + R(t) = \text{constant}, \tag{11.10.53}$$

$$N_2(t) = \bar{S}(t) + \bar{I}(t) + \bar{R}(t) = \text{constant}. \tag{11.10.54}$$

Also, be aware that two time scales exist for this system

$$T_1 = \frac{1}{a_1}, \quad T_2 = \frac{1}{a_2}, \tag{11.10.55}$$

and, as a consequence, the denominator functions are given by the expressions

$$\phi_1(h, a_1) = \frac{1 - e^{-a_1 h}}{a_1}, \quad \phi_2(h, a_2) = \frac{1 - e^{-a_2 h}}{a_2}. \tag{11.10.56}$$

Thus, the NSFD schemes are

$$\frac{S_{k+1} - S_k}{\phi_1} = -r_1 S_{k+1} \bar{I}_k, \tag{11.10.57}$$

$$\frac{I_{k+1} - I_k}{\phi_1} = r_1 S_{k+1} \bar{I}_k - a_1 I_k, \tag{11.10.58}$$

$$\frac{R_{k+1} - R_k}{\phi_1} = a_1 I_k, \tag{11.10.59}$$

and

$$\frac{\bar{S}_{k+1} - \bar{S}_k}{\phi_2} = -r_2 \bar{S}_{k+1} I_k, \tag{11.10.60}$$

$$\frac{\bar{I}_{k+1} - \bar{I}_k}{\phi_2} = r_2 \bar{S}_{k+1} I_k - a_2 \bar{I}_k, \tag{11.10.61}$$

$$\frac{\bar{R}_{k+1} - \bar{R}_k}{\phi_2} = a_2 \bar{I}_k. \tag{11.10.62}$$

Finally, solving for the variables evaluated at the $(k + 1)$th discrete-time level gives

$$S_{k+1} = \frac{S_k}{1 + (r_1 \phi_1) \bar{I}_k}, \tag{11.10.63}$$

$$I_{k+1} = (e^{-a_1 h}) I_k + (r_1 \phi_1) S_{k+1} \bar{I}_k, \tag{11.10.64}$$

$$R_{k+1} = R_k + (a_1 \phi_1) I_k, \tag{11.10.65}$$

and

$$\bar{S}_{k+1} = \frac{S_k}{1 + (r_2 \phi_2) I_k}, \tag{11.10.66}$$

$$\bar{I}_{k+1} = (e^{-a_2 h}) \bar{I}_k + (r_2 \phi_2) \bar{S}_{k+1} I_k, \tag{11.10.67}$$

$$\bar{R}_{k+1} = \bar{R}_k + (a_2 \phi_2) \bar{I}_k. \tag{11.10.68}$$

11.10.6 *Brauer-van den Driessche SIR Model*

This model takes the form (see Mickens [3], Section 3.2)

$$\frac{dS}{dt} = d_1(m_s - S) - \beta SI + \gamma I, \tag{11.10.69}$$

$$\frac{dI}{dt} = d_1(m_i - I) + \beta SI - \gamma SI - \gamma I - \alpha I, \tag{11.10.70}$$

$$\frac{dR}{dt} = -d_1 R + \alpha I, \tag{11.10.71}$$

where $(d_1, \alpha, \beta, \gamma, m_2, m_i)$ are non-negative parameters. Defining M to be

$$M(t) = S(t) + I(t) + R(t), \tag{11.10.72}$$

it is easy to show that $M(t)$ satisfies the conservation law

$$\frac{dM}{dt} = d_1[(m_s + m_i) - M], \tag{11.10.73}$$

and has the exact finite difference scheme

$$\frac{M_{k+1} - M_k}{\phi(d_1, h)} = d_1[(m_s + m_i) - M_{k+1}]. \tag{11.10.74}$$

Requiring the positivity conditions to hold, a NSFD scheme for Eqs. (11.10.69) to (11.10.71) is

$$\frac{S_{k+1} - S_k}{\phi} = d_1(m_s - S_{k+1}) - \beta S_{k+1} I_k + \gamma I_{k+1}, \tag{11.10.75}$$

$$\frac{I_{k+1} - I_k}{\phi} = d_1(m_i - I_{k+1}) + \beta S_{k+1} I_k - \gamma I_{k+1} - \alpha I_{k+1}, \tag{11.10.76}$$

$$\frac{R_{k+1} - R_k}{\phi} = -d_1 R_{k+1} + \alpha I_{k+1}. \tag{11.10.77}$$

Inspection of these equations shows that they are linear in $(S_{k+1}, I_{k+1}, R_{k+1})$, and solving for them gives the expressions

$$M_{k+1} = \frac{M_k + d_1(m_s + m_i)}{1 + d_1 \phi}, \tag{11.10.78}$$

$$\phi(d_1, h) = \frac{1 - e^{-d_1 h}}{d_1}, \tag{11.10.79}$$

and

$$S_{k+1} = \frac{B_2 C_1 + B_1 C_2}{A_1 B_2 - A_2 B_1} = F(S_k, I_k), \tag{11.10.80}$$

$$I_{k+1} = \frac{A_2 C_1 + A_1 C_2}{A_1 B_2 - A_2 B_1} = G(S_k, I_k), \tag{11.10.81}$$

$$R_{k+1} = \frac{R_k + (\alpha\phi) I_{k+1}}{A_1 B_2 - A_2 B_1} = GH(I_{k+1}, R_k), \tag{11.10.82}$$

where

$$A_1 = (1 + d_1\phi) + (\beta\phi)I_k, \quad A_2 = (\beta\phi)I_k, \tag{11.10.83}$$

$$B_1 = \gamma\phi, \quad B_2 = (1 + d_1\phi) + (\alpha + \gamma)\phi, \tag{11.10.84}$$

$$C_1 = S_k + d_1 m_s \phi, \quad C_2 = I_k + d_1 m_i \phi, \tag{11.10.85}$$

with

$$A_1 B_2 - A_2 B_1 = (1 + d_1\phi)^2 + (1 + d_1\phi)(\alpha + \gamma)\phi$$
$$+ \alpha\beta\phi^2 + (\beta\phi)(1 + d_1\phi)I_k > 0. \tag{11.10.86}$$

In summary, a NSFD scheme has been constructed for this particular SIR model. All solutions for (S_k, I_k, R_k) are non-negative and bounded, and the discrete version of the associated conservation law is exactly satisfied.

11.10.7   *Spatial Spread of Rabies*

This model is discussed in Murray [13], Section 20.3. It considers an SIR model having diffusion taking place in the infective population. The systems three equations are

$$\frac{\partial S}{\partial t} = -rIS, \tag{11.10.87}$$

$$\frac{\partial I}{\partial t} = rIS - aI + D\frac{\partial^2 I}{\partial x^2}, \tag{11.10.88}$$

$$\frac{\partial R}{\partial t} = aI, \tag{11.10.89}$$

where (r, a, D) are non-negative parameters. Note that all three variables, (S, I, R) are functions of both x and t.

Adding the equations gives

$$\frac{\partial P(x,t)}{\partial t} = D\frac{\partial^2 I(x,t)}{\partial x^2}, \tag{11.10.90}$$

where

$$P(x,t) = S(x,t) + I(x,t) + R(x,t), \tag{11.10.91}$$

is the total population at location x and time t. D is the diffusion coefficient and for $D = 0$, we have the conservation equation

$$P(t) = S(t) + I(t) + R(t) = \text{constant}, \tag{11.10.92}$$

a situation which eliminates the dependence of the dependent variables on x.

The following is a NSFD scheme for Eqs. (11.10.87) to (11.10.89):

$$\frac{S_m^{k+1} - S_m^k}{\phi} = -rI_m^k S_m^{k+1}, \qquad (11.10.93)$$

$$\frac{I_m^{k+1} - I_m^k}{\phi} = rI_m^k S_m^{k+1} - aI_m^{k+1} + D\left[\frac{I_{m+1}^k - 2I_m^k + I_{m-1}^k}{\psi}\right], \qquad (11.10.94)$$

$$\frac{R_m^{k+1} - R_m^k}{\phi} = aI_m^{k+1}, \qquad (11.10.95)$$

where for the time being

$$\phi = \phi(\Delta t), \quad \psi = \psi(\Delta t), \qquad (11.10.96)$$

are not explicitly stated, but have the properties

$$\phi(z) = z + O(z^2), \quad \psi(z) = z^2 + O(z^4). \qquad (11.10.97)$$

In the above discretized representations, we use the notation

$$\begin{cases} x \to x_m = (\Delta x)m, & m = \text{integer}; \\ t \to t_k = (\Delta t)k, & k = \text{integer}; \\ S(x,t) \to S_m^k, \text{ etc.} \end{cases} \qquad (11.10.98)$$

Solving Eqs. (11.10.93) to (11.10.95) gives the expressions

$$S_m^{k+1} = \frac{S_m^k}{1 + (r\phi)I_m^k}, \qquad (11.10.99)$$

$$I_m^{k+1} = \frac{(r\theta)I_m^k S_m^{k+1} + R(I_{m+1}^k + I_{m-1}^k + (1 - 2R)I_m^k}{1 + a\phi}, \qquad (11.10.100)$$

$$R_m^{k+1} = R_m^k (a\phi) I_m^{k+1}, \qquad (11.10.101)$$

where

$$R \equiv \frac{\phi(\Delta t)D}{\psi(\Delta x)}. \qquad (11.10.102)$$

To obtain solutions, one first calculates S_m^{k+1} from S_m^k and I_m^k; then I_m^{k+1} is calculated in terms of $(S_m^{k+1}, I_{m+1}^k, I_m^k, I_{m-1}^k)$; and, finally, R_m^{k+1} is determined from (R_m^k, I_m^{k+1}).

A way to enforce the positivity condition, i.e.,

$$(S_m^k \geq 0, I_m^k \geq 0, R_m^k \geq 0)$$
$$\Rightarrow (S_m^{k+1} \geq 0, I_m^{k+1} \geq 0, R_m^{k+1} \geq 0), \qquad (11.10.103)$$

is to require

$$1 - 2R \geq 0. \tag{11.10.104}$$

A choice that can be made is to use

$$1 - 2R = \gamma R \quad \gamma \geq 0, \tag{11.10.105}$$

where γ is a real number and select $\phi(\Delta t)$ to be

$$\phi(\Delta t) = \frac{e^{a\Delta t} - 1}{a}. \tag{11.10.106}$$

The simplest choice for $\psi(\Delta t)$ is

$$\psi(\Delta x) = (\Delta x)^2, \tag{11.10.107}$$

and we make this selection.

From Eqs. (11.10.102), (11.10.105) and (11.10.106), it follows that

$$\Delta t = \left(\frac{1}{a}\right) \text{Ln} \left[1 + \frac{a(\Delta x)^2}{(2 + \gamma)D}\right]. \tag{11.10.108}$$

This last expression provides a functional relationship between the space and time step-sizes. Such relations are generally expected for the discretization of partial differential equations [8].

With this in mind, to calculate a numerical solution, the following procedure should be carried out:

(i) First, select the parameters (r, a, D).
(ii) Second, choose values for $\gamma \geq 0$ and Δx, and calculate Δt using Eq. (11.10.108).
(iii) Third, calculate $(S_m^{k+1}, I_m^{k+1}, R_m^{k+1})$ using the procedure given immediately after Eq. (11.10.102).

11.10.8   *Fisher Equation*

The logistic equation

$$\frac{du(t)}{dt} = u(t)[1 - u(t)], \tag{11.10.109}$$

in the above normalized form, is one of several standard single population models [7]. The Fisher equation

$$\frac{\partial u}{\partial t} = \frac{\partial^2 u}{\partial x^2} + u(1 - u), \quad u = u(x, t), \tag{11.10.110}$$

can then be considered a logistic population system with diffusion.

The solutions of interest satisfy both positivity and boundedness conditions, i.e.,

$$0 \le u(x,0) \le 1 \Rightarrow 0 \le u(x,t) \le 1, \quad t > 0. \tag{11.10.111}$$

We seek to construct such a NSFD scheme for Eq. (11.10.110). Using the discretizations listed in Eq. (11.10.98), we consider the following expression

$$\frac{u_m^{k+1} - u_m^k}{\Delta t} = \frac{u_{m+1}^k - 2u_m^k + u_{m-1}^k}{(\Delta x)^2}$$
$$+ (2\bar{u}_m^k - u_m^{k+1}) - \bar{u}_m^k u_m^{k+1}, \tag{11.10.112}$$

where

$$\bar{u}_m^k \equiv \frac{u_m^k + u_m^k + u_{m-1}^k}{3}. \tag{11.10.113}$$

While the discretizations for u_t and u_{xx} are the standard forward-Euler and central difference representations, the other terms have the nonlocal forms

$$u = 2u - u \to 2\bar{u}_m^k - u_m^{k+1}, \tag{11.10.114}$$
$$u^2 \to \bar{u}_m^k u_m^{k+1}. \tag{11.10.115}$$

Since Eq. (11.10.112) is linear in u_m^{k+1}, we can solve for it to obtain

$$u_m^{k+1} = \frac{(1 - 2R)u_m^k + R(u_{m+1}^k + u_{m-1}^k) + (2\Delta t)\bar{u}_m^k}{(1 + \Delta t) + (\Delta t)\bar{u}_m^k} \tag{11.10.116}$$

where

$$R = \frac{\Delta t}{(\Delta x)^2}. \tag{11.10.117}$$

Imposing the requirement

$$1 - 2R = \gamma R, \quad \gamma \ge 0, \tag{11.10.118}$$

to ensure positivity, we find

$$R = \frac{1}{2 + \gamma} \Rightarrow \Delta t = \frac{(\Delta x)^2}{2 + \gamma}. \tag{11.10.119}$$

To show the boundedness properties of this scheme, we note the following inequalities:

$$0 \le u_m^k \le 1, \quad \text{(assumed to be true for all } m\text{)}; \tag{11.10.120}$$

$$\frac{u_{m+1}^k + u_{m-1}^k}{2 + \gamma} \le \frac{2}{2 + \gamma}; \tag{11.10.121}$$

$$(2\Delta t)\bar{u}_m^k = (\Delta t)\bar{u}_m^k + (\Delta t)\bar{u}_m^k$$
$$\leq \Delta t + (\Delta t)\bar{u}_m^k. \tag{11.10.122}$$

If R from Eqs. (11.10.119) is substituted into the numerator of Eq. (11.10.116) and the above inequalities are used, then the following result is obtained

$$\left(\frac{\gamma}{2+\gamma}\right)u_m^k + \left(\frac{1}{2+\gamma}\right)(u_{m+1}^k + u_{m-1}^k)$$
$$+ (2\Delta t)\bar{u}_m^k \leq (1 + \Delta t) + (\Delta t)\bar{u}_m^k. \tag{11.10.123}$$

Dividing both sides of this inequality by the expression on the right-side and noting that now the left-side of the resulting expression is just u_m^{k+1}, see Eq. (11.10.116), we can conclude that

$$0 \leq u_m^{k+1} \leq 1. \tag{11.10.124}$$

This demonstrates that the proposed NSFD scheme satisfies both the boundedness and positivity requirements.

Inspection of Eq. (11.10.116) shows the appearance of the term $(1+\Delta t)$ in its denominator. Since

$$1 + \Delta t = e^{\Delta t} + O[(\Delta x)^2], \tag{11.10.125}$$

we can make the replacement [6]

$$h \to e^{\Delta t} - 1. \tag{11.10.126}$$

Therefore, the discrete time derivative is

$$u_t \to \frac{u_m^{k+1} - u_m^k}{(e^{\Delta t} - 1)}, \tag{11.10.127}$$

and making this substitution into Eq. (11.10.119) and solving for Δt gives

$$\Delta t = \text{Ln}\left[1 + \frac{(\Delta x)^2}{(2+\gamma)}\right], \tag{11.10.128}$$

and

$$R \equiv \frac{\Delta t}{(\Delta x)^2} = \left[\frac{1}{(\Delta x)^2}\right]\text{Ln}\left[1 + \frac{(\Delta x)^2}{(2+\gamma)}\right]. \tag{11.10.129}$$

References

1. H. Goldstein, *Classical Mechanics*, 2nd edition (Addison-Wesley, London, 1980).
2. R. E. Mickens, *Contemporary Mathematics* **410**, 279–296 (2006). Application of NSFD methods to the numerical integration of biosciences differential equation models.
3. R. E. Mickens, *Journal of Biological Dynamics* **1**, 427–436 (2007). Numerical integration of population models satisfying conservation laws: NSFD methods.
4. R. E. Mickens and T. M. Washington, *Computers and Mathematics with Applications* **66**, 2307–2316 (2013). NSFD discretizations of interacting population models satisfying conservation laws.
5. R. E. Mickens, *Journal of Difference Equations and Applications* **11**, 645–653 (2005). Dynamic consistency: a fundamental principal for constructing NSFD schemes for differential equations.
6. R. E. Mickens, *Numerical Methods for Partial Differential Equations* **23**, 672–691 (2007). Calculation of denominator for nonstandard finite difference schemes for differential equations for satisfying a positivity condition.
7. J. D. Murray, *Mathematical Biology* (Springer-Verlag, Berlin, 1989). See Section 19.2.
8. R. E. Mickens, *Nonstandard Finite Difference Models of Differential Equations* (World Scientific, Singapore, 1994).

11.11 Black-Scholes Equations

The so-called Black-Scholes partial differential equations have played important roles in the modelling and understanding of certain financial markets [1, 2, 3]. Mathematically, these equations take the form [4]

$$\frac{\partial U}{\partial t} = Dx^2 \frac{\partial^2 U}{\partial x^2} + x \frac{\partial U}{\partial x} - rU, \quad U = U(x,t), \qquad (11.11.1)$$

or, the equivalent equation [4]

$$\frac{\partial U}{\partial t} = D \frac{\partial^2 U}{\partial x^2} + (r - D) \frac{\partial U}{\partial x} - rU. \qquad (11.11.2)$$

(Note that the x variables in these two equations are not the same variable; they are related to each other by exponential transformation. See Mickens [4].)

Exact finite difference schemes have been derived for the t-independent versions of Eqs. (11.11.1) and (11.11.2), and this construction appears in Section 11.6. Our purpose here is to construct NSFD discretizations for the full PDE's. The method of subequations will play an important role in this procedure.

To begin, note that both Eqs. (11.11.1) and (11.11.2) have the subequation [6]

$$\frac{\partial U}{\partial t} = -rU, \tag{11.11.3}$$

and this linear, first-order ODE has the exact scheme

$$\frac{U^{k+1} - U^k}{\phi(\Delta t, r)} = -rU^{k+1}, \quad \phi(\Delta t, r) = \frac{e^{r\Delta t} - 1}{r}, \tag{11.11.4}$$

where

$$t_k = (\Delta t)k; \quad k = 0, 1, 2, \ldots; \quad U^k = U(t_k). \tag{11.11.5}$$

If we now match this term with the corresponding term for the exact finite difference schemes for the t-independent versions of Eqs. (11.11.1) and (11.11.2), then the following results are found

Eq. (11.11.1):

$$\frac{U_m^{k+1} - U_m^k}{\phi(\Delta t, r)} = r \left(\frac{A_m}{B_m - A_m - C_m} \right) (u_{m+2}^k - 2U_{m+1}^k + U_m^k)$$
$$+ r \left(\frac{2A_m - B_m}{B_m - A_m - C_m} \right) (u_{m+1}^k - U_m^k)$$
$$- rU_m^{k+1}; \tag{11.11.6}$$

Eq. (11.11.2):

$$\frac{U_m^{k+1} - U_m^k}{\phi(\Delta t, r)} = \left(\frac{-r}{G_2} \right) \Delta^2 U_m^k + \left(\frac{-G_1 r}{G_2} \right) \Delta U_m^k - rU_m^{k+1}, \tag{11.11.7}$$

where

$$\Delta f_m^k \equiv f_{m+1}^k - f_m^k. \tag{11.11.8}$$

The functions (A_m, B_m, C_m) and (G_1, G_2) are given in Section 11.6.

With a little algebraic effort, Eqs. (11.11.6) and (11.11.7) can be transformed to the expressions

Eq. (11.11.6):

$$U_m^{k+1} = (e^{-r\Delta t})U_m^k \tag{11.11.9}$$

$$+ \left(\frac{1 - e^{-r\Delta t}}{B_m - A_m - C_m} \right) [A_m U_{m+2}^k - B_m U_{m+1}^k + (B_m - A_m)U_m^k], \tag{11.11.10}$$

Eq. (11.11.7):

$$U_m^{k+1} = (e^{-r\Delta t})U_m^k - \left(\frac{1 - e^{-r\Delta t}}{G_2} \right) U_{m+2}^k \tag{11.11.11}$$

$$- (2 + G_1)U_{m+1}^k + (1 + G_1)U_m^k. \tag{11.11.12}$$

Examination of Eqs. (11.11.9) and (11.11.11) allows the following conclusions to be reached:

(a) both NSFD schemes are explicit;
(b) to have positivity, restrictions will have to be placed on Δx, and a functional relationship will exist between Δt and Δx.

Finally, it should be indicated that the methodology of this section has been applied to a generalized, nonlinear version of the Black-Scholes pricing PDE [5]; the equation considered is (in our notation)

$$\frac{\partial U}{\partial t} = \frac{\sigma^2 x^2}{2 \left[1 - \lambda(x, t)\frac{\partial^2 U}{\partial x^2} \right]^2} \cdot \frac{\partial^2 U}{\partial x^2} + rx\frac{\partial U}{\partial x} - rU, \tag{11.11.13}$$

where $\lambda(x, t)$ is a specified function. Their scheme uses an exact scheme for the linear convection-reaction terms and the space derivative is approximated by a NSFD scheme. They show that the proposed discretization preserves positivity, as well as stability and consistency.

References

1. F. Black and M. Scholes, *Journal of Political Economics* **81**, 637–659 (1973). The pricing of options and corporate liabilities.
2. R. C. Merton, *Bell Journal of Economics and Management Sciences* **4**, 141–183 (1973). Theory of rational option pricing.
3. P. Wilmott, S. Howison, and J. Dewynne, *The Mathematics of Financial Derivatives* (Cambridge University Press, New York, 2002).

4. R. E. Mickens, J. Munyakazi, and T. M. Washington, *Journal of Difference Equations and Applications*, (2015). DOI: 10.1080/10236198.2015.1034118. A note on the exact disretization for a Cauchy-Euler equation: application to the Black-Scholes equation.

5. A. J. Arenas, G. G. Gonzalez-Parra, and Blas Melendex Caraballo, *Mathematical and Computer Modelling* **57**, 1663–1670 (2013). A nonstandard finite difference scheme for a nonlinear Black-Scholes equation.

6. R. E. Mickens, *Nonstandard Finite Difference Models of Differential Equations* (World Scientific, Singapore, 1994).

11.12   Time-Independent Schrödinger Equations

The time-independent Schrödinger equations [1] provide a powerful mathematical model for a broad range of phenomena in the areas of atomic, nuclear, ocean acoustics, and optical physics; see Merzbacher [1] and other references in Mickens, Chapter 8 [2]. This second-order ODE is

$$-\left(\frac{\hbar^2}{2m}\right)\frac{d^2\phi(x)}{dx^2} + V(x)\phi(x) = E\phi(x), \tag{11.12.1}$$

where in atomic physics, $V(x)$ is the potential energy function, E is the total energy, m is the electron mass, and $\hbar = h/2\pi$ is the scaled Planck constant [1]. This equation can be rewritten to the form

$$\frac{d^2\phi(x)}{dx^2} = -W(x)\phi(x), \tag{11.12.2}$$

where in atomic units, $\hbar = m = 1$, an

$$W_x = 2(E - V(x)). \tag{11.12.3}$$

The function $\phi(x)$ is normalized, i.e.,

$$\int_{-\infty}^{\infty} [\phi(x)]^2 dx = 1, \tag{11.12.4}$$

and satisfies the boundary conditions

$$\phi(-\infty) = 0, \quad \phi(+\infty) = 0. \tag{11.12.5}$$

For a given potential energy function, $V(x)$, the goal is to determine the discrete eigenvalues for the energy, E. A way to do this is to numerically solve the general time-dependent Schrödinger equation using imaginary time, i.e., [3]

$$\frac{\partial\psi(x,t)}{\partial t} = \left(\frac{1}{2}\right)\frac{\partial^2\psi(x,t)}{\partial x^2} - V(x)\psi(x,t), \tag{11.12.6}$$

which is a diffusion type PDE. The procedure is to start with a proper initial wave function, $\psi(x,0)$, and have it evolve to $\phi_0(x)$, the ground state wavefunction. Of course, the method will work only if in fact [3, 4]

$$\text{Lim}_{t\to\infty} \psi(x,t) = \phi_0(x). \tag{11.12.7}$$

To obtain $\phi_1(x)$, the wavefunction for the first excited state, the initial guess for $\psi(x,0)$ must be orthogonal to that of the initial selection to determine $\phi_0(x)$.

After the wave functions are obtained (numerically), the corresponding energies are computed numerically from the relation

$$E = \frac{\int_{-\infty}^{\infty} \phi^*(x)\bar{H}\phi(x)dx}{\int_{-\infty}^{\infty} |\phi(x)|^2 dx}, \tag{11.12.8}$$

where $\bar{H}$ is a discretized form of the operator

$$\hat{H} = -\left(\frac{1}{2}\right)\frac{\partial^2}{\partial x^2} + V(x). \tag{11.12.9}$$

The paper by Angraini and Sudiarta [5] discusses four possible finite difference schemes for the discretization of the second-derivative in Eq. (11.2.6). If we ignore the time discretization, and denote the wavefunction by $\psi(x)$, we have:

Standard Central Difference

$$\left[\frac{\partial^2 \psi(x)}{\partial x^2}\right]_{x=(\Delta x)m} \to \frac{\psi_{m+1} - 2\psi_m + \psi_{m-1}}{(\Delta x)^2}; \tag{11.12.10}$$

Mickens NSFD Scheme

$$\left[\frac{\partial^2 \psi(x)}{\partial x^2}\right]_{x=(\Delta x)m} \to \frac{\psi_{m+1} - 2\psi_m + \psi_{m-1}}{g(x_m)}, \tag{11.12.11}$$

where

$$g(x_m) = \begin{cases} (\Delta x)^2, & W(x_m) = 0; \\ \left(\frac{4}{W_m}\right)\sin^2\left[\sqrt{W_m}\left(\frac{\Delta x}{2}\right)\right], & W(x_m) > 0; \\ \left(\frac{4}{|W_m|}\right)\sinh^2\left[\sqrt{|W_m|}\left(\frac{\Delta x}{2}\right)\right], & W(x_m) < 0; \end{cases} \tag{11.12.12}$$

where $W_m = W(x_m)$.

Numerov Scheme

$$
\left[\frac{\partial^2 \psi(x)}{\partial x^2}\right]_{x=(\Delta x)m} \to \left\{\left[1 - \frac{(\Delta x)^2}{12}W_{m+1}\right]\psi_{m+1}\right.
$$
$$
- 2\left[1 + \frac{(\Delta x)^2}{12}W_m\right]\psi_m
$$
$$
\left. + \left[1 - \frac{(\Delta x)^2}{12}W_{m-1}\right]\psi_{m-1}\right\} \Bigg/ (\Delta x)^2; \quad (11.12.13)
$$

NSFD-Numerov Scheme

$$
\left[\frac{\partial^2 \psi(x)}{\partial x^2}\right]_{x=(\Delta x)m} \to \left\{\left[1 - \frac{(\Delta x)^2}{12}W_{m+1}\right]\psi_{m+1}\right.
$$
$$
- 2\left[1 + \frac{(\Delta x)^2}{12}W_m\right]\psi_m
$$
$$
\left. + \left[1 - \frac{(\Delta x)^2}{12}W_{m-1}\right]\psi_{m-1}\right\} \Bigg/ g_m\left[1 + \frac{(\Delta x)^2}{12}W_m\right].
$$
$$
(11.12.14)
$$

To repeat, in the previous equations, we use the interchangeable notations

$$
\psi(x_m) = \psi_m, \quad g(x_m) = g_m, \quad (11.12.15)
$$

where g_m is taken from Eq. (11.12.12).

Given ψ_m, the energy of the system is numerically calculated by using the formula [5]

$$
E = \left[\frac{1}{\sum_m (\psi_m)^2}\right]\sum_m\left\{\left(-\frac{1}{2}\right)\psi_m\left[\frac{\partial^2 \psi(x)}{\partial x^2}\right]_{x=(\Delta x_m} + V_m[\psi_m]^2\right\}
$$
$$
(11.12.16)
$$

where $V_m = V(x_m)$.

This methodology was tested on four standard potentials appearing in quantum mechanics, namely,

- infinite square well
- harmonic oscillator
- symmetric Poschl-Teller
- Morse

and the following results were obtained [5]:

> "Three finite difference schemes were used to modify the standard finite difference time domain (FDTD) method. It has been shown that the non-standard FDTD and Numerov-NSFDTD gave generally more accurate results when compared to the standard FDTD method. The Numerov-NSFDTD method was generally shown to perform better than the standard FDTD method for all cases."

References

1. E. Merzbacher, *Quantum Mechanics* (Wiley, New York, 1961).
2. R. E. Mickens, *Nonstandard Finite Difference Models of Differential Equations* (World Scientific, Singapore, 1994). See Chapter 8.
3. I. W. Sudiarta and D. W. Geldart, *Journal of Physics A: Mathematical and Theoretical* **40**, 1885–1891 (2007). Solving the Schrödinger equation using the finite difference time domain method.
4. I. W. Sudiarta, *Pramana* **91**, 52–58 (2018). Non-standard finite-difference time-domain method for solving the Schrödinger equation.
5. L. M. Angraini and I. W. Sudiarta, *Journal of Physics: Theories and Applications* **2**, 27–33 (2018). Non-standard and Numerov finite difference schemes for finite difference time domain method to solve one-dimensional Schrödinger equations.

11.13   Linear, Damped Wave Equation

A general PDE modelling linear, damped, wave motion is

$$\tau\frac{\partial^2 u}{\partial t^2} - D\frac{\partial^2 u}{\partial x^2} + \gamma\frac{\partial u}{\partial t} - \beta^2 u = 0, \quad u = u(x,t), \qquad (11.13.1)$$

where all the parameters $(\tau, D, \gamma, \beta^2)$ are non-negative. For particular values of its parameters this equation represents a diverse range of phenomena in the natural and engineering sciences [1, 2]. The case for which $\beta^2 = 0$ has been studied extensively by several authors who examined various issues related to constructing finite difference discretizations such that the resulting numerical solutions are non-negative [3, 4, 5, 6]. Additional investigations considered the inclusion of nonlinear reaction term [7]. Our purpose in this section is to summarize the related work on Eq. (11.13.1) as it relates to a finite difference scheme having solutions which are non-negative. The results to be presented summarize that achieved by Macias-Diaz and Gallegos [7].

After scaling of both dependent and independent variables, Eq. (11.13.1) becomes equivalent to the expression

$$\tau \frac{\partial^2 u}{\partial t^2} - \frac{\partial^2 u}{\partial x^2} + \frac{\partial u}{\partial t} - \beta^2 u = 0, \tag{11.13.2}$$

where (u, x, t) are the rescaled variables. The NSFD discretization is constructed by making the replacement

$$u_{tt} \rightarrow \frac{u_m^{k+1} - 2u_m^k + u_m^{k-1}}{(\Delta t)^2}, \tag{11.13.3}$$

$$u_{xx} \rightarrow \frac{u_{m+1}^k - (u_m^{k+1} + u_m^{k-1}) + u_{m-1}^k}{(\Delta x)^2}, \tag{11.13.4}$$

$$u_t \rightarrow (1 - \alpha) \left[\frac{u_m^{k+1} - u_m^k}{\Delta t} \right] + \alpha \left[\frac{u_m^{k+1} - u_m^{k-1}}{2\Delta t} \right], \tag{11.13.5}$$

$$u \rightarrow \frac{u_{m+1}^k + u_{m-1}^k}{2}, \tag{11.13.6}$$

where, for the moment, α is an undetermined parameter. Substituting these expressions into Eq. (11.13.2) and simplifying gives

$$k_1 u_m^{k+1} = k_2 u_m^k + k_3 u_m^{k-1} + k_4 (u_{m+1}^k + u_{m-1}^k), \tag{11.13.7}$$

where (k_1, k_2, k_3, k_4) depend on $(\Delta x, \Delta t)$ and (τ, β). In more detail

$$k_1 = \frac{\tau}{\Delta t} + R + 1 - \frac{\alpha}{2}, \tag{11.13.8}$$

$$k_2 = \frac{2\tau}{\Delta t} + 1 - \alpha, \tag{11.13.9}$$

$$k_3 = \frac{\alpha}{2} - \frac{\tau}{\Delta t} - R, \tag{11.13.10}$$

$$k_4 = R + \frac{\beta^2 \Delta t}{2}, \tag{11.13.11}$$

where

$$R = \frac{\Delta t}{(\Delta x)^2}. \tag{11.13.12}$$

To ensure that the positivity condition holds, i.e.,

$$u_m^{k-1} \geq 0, \quad u_m^k \geq 0 \Rightarrow u_m^{k+1} \geq 0, \tag{11.13.13}$$

we let

$$k_3 = R, \tag{11.13.14}$$

and this gives

$$\alpha = 2\left(2R + \frac{\tau}{\Delta t}\right), \tag{11.13.15}$$

$$k_1 = 1 - R, \quad k_2 = 1 - 4R. \tag{11.13.16}$$

To proceed, take

$$R = \frac{1}{4}, \tag{11.13.17}$$

and this choice allows the rewriting of Eq. (11.13.7) to the form

$$u_m^{k+1} = \left(\frac{1}{3}\right)[u_m^{k-1} + (1 + 2\beta^2 \Delta t)(u_{m+1}^k + u_{m-1}^k)]. \tag{11.13.18}$$

The paper of Macias-Diaz and Gallegos [7] establish the following features for the above constructed NSFD scheme for Eq. (11.13.18):

(i) For $R = \frac{1}{4}$, the scheme has a unique solution for bounded initial conditions.

(ii) The scheme is conditional stable, i.e., this holds, in general, for Eq. (11.13.7) for $R \leq \frac{1}{2}$.

(iii) The scheme, Eq. (11.13.7) is both consistent and convergent for $\Delta t \leq 1$ and $R < 1$, with a linear order of convergence in time, and a second order of convergence in space.

Numerical simulations of the scheme showed that it provides somewhat better results than a standard method. In particular, its numerical solutions were always non-negative, in contrast to a standard procedure which gave negative solutions for certain parameter values.

References

1. J. D. Logan, *An Introduction to Nonlinear Partial Differential Equations* (Wiley, New York, 2008).
2. J. D. Murray, *Mathematical Biology, Vols. I and II*, 3rd Edition (Springer Verlag, New York, 2004).
3. R. E. Mickens and P. M. Jordan, *Numerical Methods for Partial Differential Equations* **20**, 639–649 (2004). A positivity-preserving nonstandard finite difference scheme for the damped wave equation.

4. R. E. Mickens and P. M. Jordan, *Numerical Methods for Partial Differential Equations* **21**, 976–985 (2005). A new positivity-preserving NSFD scheme for the DWE.
5. J. E. Macias and A. Puri, *Applied Numerical Mathematics* **60**, 934–948 (2010). A boundedness-preserving finite-difference scheme for a damped nonlinear wave equation.
6. R. Čiegis and A. Mirinavičius, *Central European Journal of Mathematic* **9**, 1164–1170 (2011). On some finite difference schemes for solutions of hyperbolic heat conduction problems.
7. J. E. Macias-Diaz and A. Gallegos, *Journal of Computational and Applied Mathematics* **354**, 603–611 (2019). On a positivity-preserving numerical model for a linearized hyperbolic Fisher-Kolmogorov-Petrovski-Piscounov equation.

11.14   NSFD Constructions for Burgers and Burgers-Fisher PDE's

A great deal of analytical study has been devoted to the class of PDE's modelling advection, diffusion, and reaction processes [1, 2]. Similarly, the NSFD methodology has been applied to these types of equations, which include the so-called Burgers and Burgers-Fisher PDE [3, 4].

Our purpose here is to provide a summary of a particular set of results obtained by Zhang et al. [5] for the Burgers PDE

$$u_t + \alpha u u_x = u_{xx}, \quad u = u(x,t), \qquad (11.14.1)$$

and, Burgers-Fisher PDE

$$u_t + \alpha u u_x = u_{xx} + u(1-u), \qquad (11.14.2)$$

where α is, in general, a positive parameter. Their "exact schemes" are constructed from having knowledge of exact solitary wave solutions to the respective partial differential equations. They also constructed, separately, other NSFD discretizations. Below, we give a summary of what they found and request readers to read their interesting and creative article for the details.

11.14.1   *Burgers Equations: $u_t + u u_x = u_{xx}$*

Based on a discretization of the exact solitary wave solution

$$u(x,t) = \frac{1}{1 + \exp\left[\left(\frac{1}{2}\right)\left(x - \frac{t}{2}\right)\right]}, \qquad (11.14.3)$$

where

$$0 \le u(x,t) \le 1, \tag{11.14.4}$$

and the fact that if we choose

$$\Delta t = 2h, \quad h = \Delta x, \tag{11.14.5}$$

then

$$u(x + h, t) = u(x, t - \Delta t), \tag{11.14.6}$$

and from this, they construct implicit and explicit "exact schemes":

Exact Implicit Scheme: $u_t + uu_x = u_{xx}$

$$\frac{u_m^{k+1} - u_m^k}{\phi_1} + u_m^{k+1} \left[\frac{u_{m+1}^{k+1} - u_{m-1}^{k+1}}{2\psi_1} \right] = \frac{u_{m+1}^{k+1} - 2u_m^{k+1} + u_{m-1}^{k+1}}{\psi_1 \psi_2}. \tag{11.14.7}$$

Exact Explicit Scheme: $u_t + uu_x = u_{xx}$

$$\frac{u_m^{k+1} - u_m^k}{\phi_2} + u_m^k \left[\frac{u_{m+1}^k - u_{m-1}^k}{2\psi_2} \right] = \frac{u_{m+1}^k - 2u_m^k + u_{m-1}^k}{\psi_1 \psi_2}. \tag{11.14.8}$$

In both of the above expressions

$$t \to t_k = (\Delta t)k, \quad x \to x_m = hm, \quad u(x,t) \to u_m^k \tag{11.14.9}$$

and

$$\psi_1 = \frac{1 - e^{-(h/2)}}{\left(\frac{1}{2}\right)}, \quad \psi_2 = \frac{e^{(h/2)} - 1}{\left(\frac{1}{2}\right)}, \tag{11.14.10a}$$

$$\phi_1 = \frac{1 - e^{-\left(\frac{\Delta t}{4}\right)}}{\left(\frac{1}{4}\right)}, \quad \phi_2 = \frac{e^{\left(\frac{\Delta t}{4}\right)} - 1}{\left(\frac{1}{4}\right)}. \tag{11.14.10b}$$

Note that

$$\psi_1(h) = h + O(h^2), \quad \psi_2(h) = h + O(h^2) \tag{11.14.11a}$$

$$\phi_1(\Delta t) = \Delta t + O[(\Delta t)^2], \quad \phi_2(\Delta t) = \Delta t + O[(\Delta t)^2], \tag{11.14.11b}$$

$$\psi(h)\psi(h) = h^2 + O(h^3), \tag{11.14.11c}$$

and we also have

$$\Delta t = 2h. \tag{11.14.11d}$$

The corresponding NSFD scheme obtained by Zhang et al. [5] is

$$\frac{u_m^{k+1} - u_m^k}{\Phi} + u_m^{k+1}\left(\frac{u_m^k - u_{m-1}^k}{\Gamma}\right) = \frac{u_{m+1}^k - 2u_m^k + u_{m-1}^k}{\Psi}, \qquad (11.14.12)$$

where

$$\Phi(\Delta t) = \frac{1 - e^{-\left(\frac{\Delta t}{4}\right)}}{\left(\frac{1}{4}\right)}, \quad \Psi(h) = \frac{(e^{h/2} - 1)^2}{\left(\frac{1}{4}\right)}, \qquad (11.14.13a)$$

$$\Gamma(h) = \frac{e^{h/2} - 1}{\left(\frac{1}{2}\right)}. \qquad (11.14.13b)$$

Likewise,

$$\Phi(\Delta t) = \Delta t + O[(\Delta t)^2], \qquad (11.14.14a)$$

$$\Psi(h) = h^2 + O(h^3), \quad \Gamma(h) = h + O(h^2). \qquad (11.14.14b)$$

If we now define R and r as

$$R = \frac{\Phi}{\Psi}, \quad r = \frac{\Phi}{\Gamma}, \qquad (11.14.15)$$

then Eq. (11.14.12) can be rewritten as

$$u_m^{k+1} = \frac{R(u_{m+1}^k + u_{m-1}^k) + (1 - 2R)u_m^k}{1 + r(u_m^k - u_{m-1}^k)}. \qquad (11.14.16)$$

Two major expected features expected of any valid discretization of the Burgers equation are that the numerical solutions should be non-negative and bounded. The authors, in fact, prove the following result:

Theorem. *If* $1 - 2R - r > 0$, *the numerical solution of Eq. (11.14.16) satisfies*

$$0 \le u_m^k \le 1 \Rightarrow 0 \le u_m^{k+1} \le 1, \qquad (11.14.17)$$

for all relevant values of k *and* m.

11.14.2   Burgers-Fisher Equations: $u_t + uu_x = u_{xx} + u(1 - u)$

An exact solitary wave solution to the Burgers-Fisher equation is

$$u(x, t) = \frac{1}{1 + \exp\left[\left(\frac{1}{2}\right)\left(x - \frac{5t}{2}\right)\right]}, \qquad (11.14.18)$$

where

$$0 \le u(x, t) \le 1. \qquad (11.14.19)$$

For this solution $u(x + h, t) = u(x, t - \Delta x)$, provided that

$$\Delta t = \left(\frac{2}{5}\right) h. \qquad (11.14.20)$$

Using these results Zhang et al. [5] construct two "exact" finite difference discretizations.

Exact Implicit Scheme: $u_t + uu_x = u_{xx} + u(1 - u)$

$$\frac{u_m^{k+1} + u_m^k}{\phi_1} + u_m^{k+1} \left[\frac{u_{m+1}^{k+1} - u_{m-1}^{k+1}}{2\psi_1} \right]$$
$$= \frac{u_{m+1}^{k+1} - 2u_m^{k+1} + u_{m-1}^{k+1}}{\psi_1 \psi_2} + u_{m+1}^{k+1}(1 - u_m^{k+1}). \qquad (11.14.21)$$

Exact Explicit Scheme: $u_t + uu_x = u_{xx} + u(1 - u)$

$$\frac{u_m^{k+1} - u_m^k}{\phi_2} + u_m^k \left(\frac{u_{m+1}^k - u_{m-1}^k}{2\psi_2} \right)$$
$$= \frac{u_{m+1}^k - 2u_m^k + u_{m-1}^k}{\psi_1 \psi_2} + u_{m-1}^k(1 - u_m^k). \qquad (11.14.22)$$

The denominator functions are given by the expressions

$$\psi_1 = \frac{1 - e^{-\frac{h}{2}}}{\left(\frac{1}{2}\right)}, \quad \psi_2 = \frac{e^{\frac{h}{2}} - 1}{\left(\frac{1}{2}\right)}, \qquad (11.14.23a)$$

$$\phi_1 = \frac{1 - e^{\frac{-5\Delta t}{4}}}{\left(\frac{5}{4}\right)}, \quad \phi_2 = \frac{e^{\frac{5\Delta t}{4}} - 1}{\left(\frac{5}{4}\right)}, \qquad (11.14.23b)$$

with

$$\Delta t = \left(\frac{2}{5}\right) h. \qquad (11.14.23c)$$

The NSFD scheme constructed by Zhang et al. [5] is

$$\frac{u_m^{k+1} - u_m^k}{\Phi} + u_m^{k+1} \left(\frac{u_m^k - u_{m-1}^k}{\Gamma} \right)$$
$$= \frac{u_{m+1}^k - 2u_m^k + u_{m-1}^k}{\Psi} + u_m^k(1 - u_m^{k+1}), \qquad (11.14.24)$$

where (Φ, Ψ, Γ) are given in Eq. (11.14.13). Using the definitions $R = \Phi/\Psi$ and $r = \Phi/\Gamma$, we can rewrite Eq. (11.14.24) to the form

$$u_m^{k+1} = \frac{R(u_{m+1}^k + u_{m-1}^k) + (1 - 2R + \Phi)u_m^k}{1 + (u_m^k - u_{m-1}^k) + \Phi u_m^k}. \qquad (11.14.25)$$

The authors prove the following theorem:

Theorem. *If* $1 - 2R - r \geq 0$, *the solution, Eq.* (11.14.25) *satisfies*

$$0 \leq u_m^k \leq 1 \Rightarrow 0 \leq u_m^{k+1} \leq 1, \qquad (11.14.26)$$

for all relevant values of k and m.

Numerical experiments showed these schemes to be dynamical consistent with the properties of positivity and boundedness, and to also be accurate [5].

References

1. J. D. Murray, *Mathematical Biology: I. An Introduction*, 3rd Edition (Springer, New York, 2002).
2. G. B. Whitham, *Linear and Nonlinear Waves* (Wiley, New York, 1974).
3. R. E. Mickens, *Numerical Methods for Partial Differential Equations* **15**, 201–214 (1999). Nonstandard finite difference schemes for reaction-diffusion equations.
4. R. E. Mickens and A. B. Gumel, *Journal Sound and Vibration* **257** 791–797 (2002). Construction and analysis of a non-standard finite difference scheme for the Burgers-Fisher equation.
5. L. Zhang, L. Wang, and X. Ding, *Journal of Applied Mathematics*, Volume 2014, Article ID 597926, 12 pages, http://dx.doi.org/10.1155/2014/597926.

11.15   Cross-Diffusion

One way to characterize interacting populations with cross-diffusion is to state that the diffusion matrix is not strictly diagonal and may not be symmetric positive. The book by Murray [1] gives several examples of such systems. Cross-diffusion, coupled PDE's appear as models for a wide range of phenomena such as cancer growth [2, 3], population dynamics via Volterra-Lotka cross-diffusion [4, 5], and chemotaxis [6].

Systems having cross-diffusion generally give rise to NSFD discretization which have the form (for two interacting population, $u(x,t)$ and $v(x,t)$)

$$u_m^{k+1} = F_m^k - G_m^k, \qquad (11.15.1)$$

$$\underset{m}{\overset{k+1}{}} = H_m^k - I_m^k, \qquad (11.15.2)$$

where (F, G, H, I) are functions of (u^k, v^k) and have the property

$$\begin{cases} F_m^k > 0, & G_m^k > 0, & H_m^k > 0, & I_m^k > 0, \\ \text{for } u^k > 0, & v^k > 0, \end{cases} \tag{11.15.3}$$

where

$$\begin{cases} u^k = (u_{m+1}^k, u_m^k, u_{m-1}^k), \\ v^k = (v_{m+1}^k, v_m^k, v_{m-1}^k). \end{cases} \tag{11.15.4}$$

These results do not allow a guarantee that the positivity condition holds, i.e.,

$$u_m^k \geq 0, \quad v_m^k \geq 0 \Rightarrow u_m^{k+1} \geq 0, \quad v_m^{k+1} \geq 0. \tag{11.15.5}$$

In spite of the fact that this seems to indicate a breakdown of the NSFD methodology, earlier work by Mickens showed how this difficulty can be resolved [7]. We illustrate the technique using the following example:

$$\frac{dx}{dt} = -\lambda x^{1/3}, \quad \lambda > 0, \quad x(0) > 0. \tag{11.15.6}$$

A NSFD scheme can be constructed by noting that

$$x^{1/3} = x^{1/3} \cdot 1 \rightarrow x_k^{1/3} \cdot (1)_k, \tag{11.15.7}$$

where $(1)_k$ is an approximation for "1." Several particular approximations are

$$(1)_k = \begin{cases} \frac{x_{k+1}}{x_k}, \\ \frac{2x_{k+1}}{x_{k+1}+x_k}, \\ \text{etc.} \end{cases} \tag{11.15.8}$$

So, a possible NSFD scheme for Eq. (11.15.6) is

$$\frac{x_{k+1} - x_k}{h} = -\lambda x_k^{1/3} \cdot \frac{x_{k+1}}{x_k}, \tag{11.15.9}$$

or

$$\frac{x_{k+1} - x_k}{h} = -\lambda \frac{x_{k+1}}{(x_k)^{2/3}}. \tag{11.15.10}$$

Solving for x_{k+1} gives

$$x_{k+1} = \left[\frac{(x_k)^{2/3}}{\lambda h + (x_k)^{2/3}} \right] x_k. \tag{11.15.11}$$

Observe that this scheme not only satisfies the positivity condition

$$x_k > 0 \Rightarrow x_{k+1} > 0, \tag{11.15.12}$$

it also produces a numerical solution which decreases monotonic to zero.

This rather clever "trick" has been used by a number of researchers to construct positivity preserving NSFD discretizations for systems with cross-diffusion [8, 9]; see also, Songolo [10], Khalsaraei et al. [11], and Patidar and Ramanantoanina [12] for related procedures.

The most detailed application of the approximation of "1" method is Chapwanya et al. [9]. We list below the three systems they consider along with the relevant partial differential equations. Readers interested in the explicit details, including proofs of some related theorems, should examine this paper.

A Model of Malignant Invasion

For this model there are three dependent variables

$$u(x,t) \ : \text{density of invasive cells,}$$
$$c(x,t) \ : \text{density of connective tissue,}$$
$$p(x,t) \ : \text{density of protease.}$$

The corresponding three coupled, nonlinear, partial differential equations are [13]

$$\frac{\partial u}{\partial t} = u(1-u) - \frac{\partial}{\partial x}\left(u\frac{\partial c}{\partial x}\right) \qquad (11.15.13)$$

$$\frac{\partial c}{\partial t} = -pc, \qquad (11.15.14)$$

$$\frac{\partial p}{\partial t} = \left(\frac{1}{\epsilon}\right)(uc-p). \qquad (11.15.15)$$

The cross-diffusion term, $(uc_x)_x$, depends on two independent variables, u and c, and is preceded also by a minus sign. Further, the authors create two NSFD scheme based on the following approximations to "1"

$$(1)_k : \frac{2u_m^{k+1}}{u_m^{k+1} + u_m^k} \quad \text{and} \quad \frac{u_m^{k+1}}{u_m^k}. \qquad (11.15.16)$$

Convective Predator-Prey Pursuit and Evasion Model [1]

$$u(x,t) \ : \text{predator density,}$$
$$v(x,t) \ : \text{prey density.}$$

This model describes a predator-prey system in which the prey attempt to evade the predators, while the predators try to catch the prey only if they interact. The two evolution partial differential equations are

$$\frac{\partial u}{\partial t} = \frac{\partial}{\partial x}\left[\left(c_1 + h_1\frac{\partial v}{\partial x}\right)u\right] + (1 - u - v)u, \tag{11.15.17}$$

$$\frac{\partial v}{\partial t} = \frac{\partial}{\partial x}\left[\left(c_2 + h_2\frac{\partial u}{\partial x}\right)v\right] + (u - 1)v. \tag{11.15.18}$$

The parameters, (c_1, c_2, h_1, h_2) are non-negative. For the NSFD discretizations the following $(1)_k$ approximations were used in the appropriate equation

$$(1)_k : \frac{u_m^{k+1}}{u_m^k}, \quad \frac{v_m^{k+1}}{v_m^k}. \tag{11.15.19}$$

Note that the cross-diffusion terms are the second items in the diffusion terms, appearing on the right-sides of Eqs. (11.15.17) and (11.15.18).

Basic Reaction-Diffusion-Chemotaxis Model [1]

$$N(x, t) \; : \text{bacterial population density},$$
$$a(x, t) \; : \text{food that bacteria consume}.$$

The model equations are

$$\frac{\partial N}{\partial t} = \alpha\frac{\partial^2 N}{\partial x^2} - \beta\frac{\partial}{\partial x}\left(N\frac{\partial a}{\partial x}\right), \tag{11.15.20}$$

$$\frac{\partial a}{\partial t} = hN - \omega a + \gamma\frac{\partial^2 a}{\partial x^2}, \tag{11.15.21}$$

where the parameters, $(\alpha, \beta, h, \omega, \gamma)$, are positive. The second term on the right-side of the first PDE is the cross-diffusion term and the $(1)_k$ approximation used was

$$(1)_k : \frac{N_m^{k+1}}{N_m^k}. \tag{11.15.22}$$

Note, no such approximation is required for the second PDE.

Clearly, an area for active research regarding the general NSFD methodology is to investigate the generalization of the "1" approximation. Note that

$$(1)_k = \frac{x_{k+1}}{x_k} + O(\Delta t). \tag{11.15.23}$$

What is required to obtain $O((\Delta t)^2)$ instead?

References

1. J. D. Murray, *Mathematical Biology II* (Springer, New York, 2003).
2. N. Bellome, et al. (Editors), *Selected Topics in Cancer Modeling* (Springer Science and Business Media, New York, 2008), pp. 255–276.
3. T. L. Jackson and H. M. Byrne, *Mathematical Biosciences* **180**, 307–328 (2002). A mechanical model of tumor encapsulation and transcapsular spread.
4. L. Chen and A. Jüngel, *Journal of Differential Equations* **224**, 39–59 (2006). Analysis of a parabolic cross-diffusion population model without self-diffusion.
5. N. Shigesada, K. Kawasaki and E. Teramoto, *Journal of Theoretical Biology* **79**, 83–99 (1979). Spatial segregation of interacting species.
6. D. Le, *SIAM Journal of Mathematical Analysis* **32**, 504–521 (2000). Coexistence with chemotaxis.
7. R. E. Mickens, Modeling "1" in the NSFD Methodology (unpublished, 2001).
8. D. T. Dimitrov and H. Kojouharov, *Journal of Difference Equations and Applications* **17**, 1721–1736 (2011). Dynamically consistent numerical methods for general production-destructive systems.
9. M. Chapwanya, J. M.-S. Lubuma and R. E. Mickens, *Computers and Mathematics with Applications* **68**, 1071–1082 (2014). Positivity-preserving nonstandard finite difference schemes for cross-diffusion equations in biosciences.
10. M. E. Songolo, *American Scientific Research Journal for Engineering, Technology, and Science* (ASRJETS), ISSN (Online) 2313–4402. A positivity-preserving nonstandard finite difference scheme for parabolic system with cross-diffusion equations and nonlocal initial conditions.
11. M. Khalsaraei, Sh. Heydari and L. D. Algoo, *Journal of Cancer Treatment and Research* **4**, 27–33 (2017). Positivity preserving nonstandard finite difference schemes applied to a cancer growth model.
12. K. C. Patidar and A. Ramanantoanina, *AIP Conference Proceedings* **2116**, 450044 (2009); https://doi.org/10.1063/1.5114511.
13. B. P. Marchant, J. Norburg, and A. J. Perumpaniani, *SIAM Journal of Applied Mathematics* **60**, 463–476 (2000). Traveling shock waves arising in a model of malignant invasion.

11.16   Delay Differential Equations

Some of the most interesting and important systems can be modeled by delay differential equations [1, 2]. These equations are a type of differential equations in which the derivative of the unknown function at a particular time is expressed in terms of the values of the function at other, generally, previous times [3].

Perhaps the simplest delay differential equation (DDE) is

$$\frac{dx(t)}{dt} = -\lambda x(t - \tau), \quad \lambda > 0, \tag{11.16.1}$$

where the constant τ is positive and is called the delay. For the case where $\tau = 0$, the assumption of a solution of the form

$$x_{\tau=0}(t) = e^{rt} \tag{11.16.2}$$

gives the characteristic equation (CE)

$$r = -\lambda, \tag{11.16.3}$$

and

$$x_{\tau=0}(t) = e^{-\lambda t}. \tag{11.16.4}$$

However, for $\tau > 0$, we have the following CE

$$r = -\lambda e^{-\tau r}, \tag{11.16.5}$$

which has an infinite number of complex-valued solutions [2, 3]. Thus, we quickly realize that DDE's are much more difficult to analyze than differential equations.

An essential question with regard to the NSFD methodology is: "Can this technique be applied to DDE's?" Fortunately, the answer is yes. The references listed at the end of this section [4–10] provide a rich sampling of some of the work done on this issue. The reader is referred to these articles both to see the broad range of DDE's which appear in the literature and also to learn the variety of techniques used to construct valid NSFD discretizations. Examination of the systems appearing in these articles demonstrate the value of DDE's to the modelling of phenomena in the biological related sciences.

For the remainder of this section, we list and briefly present information on four systems covered in the references. No explicit details of the associated NSD schemes are given. The articles should be consulted for this information.

SIRS Epidemic Model [4, 11]

$$\frac{dS(t)}{dt} = \lambda - \mu_1 S(t) - \beta(I)S(t) \int_0^\omega I(t-s)d\eta(s) + \delta R(t), \qquad (11.16.6)$$

$$\frac{dI(t)}{dt} = \beta(I)S(t) \int_0^\omega I(t-s)d\eta(s) - (\mu_2 + \gamma)I(t).I(t-s)d\eta(s) + \delta R(t),$$
$$(11.16.7)$$

$$\frac{dR(t)}{dt} = \gamma I(t) - (\mu_3 + \gamma)R(t). \qquad (11.16.8)$$

In these equations, the seven parameters

$$(\lambda, \mu_1, \mu_2, \mu_3, \gamma \delta \omega)$$

are non-negative and $\beta(I) = \beta(I(t))$ is a specified function of $I(t)$.

Class of First-Order DDE's [6, 12, 13]

$$\frac{dx(t)}{dt} = -\gamma x(t) + \beta f(x(t-\tau)), \quad t \geq 0, \qquad (11.16.9)$$

where the parameters γ and β are non-negative, and $\tau > 0$ is the time delay. The members of this class of DE's are determined by the functions $f(z)$.

Limit Cycle Oscillator [7]

Define the complex variable z as

$$z = x + iy, \quad i = \sqrt{-1}, \qquad (11.16.10)$$

where x and y are real variables. The following DDE models an "autonomously drived single limit cycle oscillator" [7]

$$\frac{dz(t)}{dt} - [a + i\omega - |z(t)|^2]z(t) = -k_1 z(t-\tau) - k_2 z^2(t-\tau). \qquad (11.16.11)$$

In the above, ω is the frequency of the oscillation, a is a real constant, k_1 and k_2 are real, and τ is the time delay.

Red Blood Cell Survival Model [9, 14]

A model to describe the survival of animal red blood cells is the following DDE [14]

$$\frac{dx(t)}{dt} = -ax(t) + b\exp[-cx(t-\tau)], \qquad (11.16.12)$$

where $a\epsilon(0,1)$, and $(b,c,\tau)\epsilon(0,\infty)$.

Comments

(a) For DDE's modelling populations, particle numbers or densities, the NSFD discretizations must be constructed such that they produce solutions that are non-negative, i.e., a positivity condition holds for the solutions.

(b) Likewise, for DDE's modelling oscillatory systems, the NSFD schemes must yield corresponding oscillatory numerical solutions.

(c) Further, if bifurcations occur for the DDE's, then they should also exist for the numerical solutions.

In other words, the discrete-time NSFD realizations must be dynamically consistent with the original continuous time DDE's.

In summary, each of the referenced publications [4–10] presents a mathematical analysis of their particular equations of interest and, additionally, include the validity of their proposed NSFD schemes. The overall conclusion is that NSFD discretizations perform better than standard methods.

References

1. R. Bellman and K. L. Cooke, *Differential-Difference Equations* (Academic Press, New York-London, 1964).
2. R. D. Driver, *Ordinary and Delay Differential Equations* (Springer-Verlag, New York, 1977).
3. https://en.wikipedia.org/wiki/Delay_differential_equation
4. M. Sekiguchi and E. Ishiwata, *Journal of Mathematical Analysis and Applications* **371**, 195–202 (2010). Global dynamics of a discretized SIRS epidemic model with time delay.
5. Y. Wang, *Communications in Nonlinear, Science and Numerical Simulation* **17**, 3967–3978 (2012). Dynamics of a nonstandard finite difference scheme for delay differential equations with unimodal feedback.

6. H. Su, L. Wenxue and X. Ding, *Journal of Mathematical Analysis and Applications* **400**, 25–34 (2013). Numerical dynamics of a nonstandard finite difference method for a class of delay differential equations.

7. Y. Wang and X. Ding, *Journal of Applied Mathematics*, Volume 2013, Article ID 912374, 11 pages (https//dx.doi.org/10.1155/2013/912374). Dynamics of a nonstandard finite-difference scheme for a limit cycle oscillates with delayed feedback.

8. S. M. Garba, A. Gumel, A. S. Hassan, and J. M.-S. Labuma, *Applied Mathematics and Computation* **258**, 388–403 (2015). Switching from exact scheme to non-standard finite difference scheme for linear delay differential equation.

9. Y. Wang, *Advances in Difference Equations* **86** (2015). DOI 10.1186/s13662-015-0432-8. Numerical dynamics of a nonstandard finite-difference-θ method for a red blood cell survival model.

10. X. Zhuang, Q. Wang and J. Wen, *International Journal of Bifurcation and Chaos* **28**, No. 11, 1850133 (2018). Numerical dynamics of NSFD method for nonlinear delay differential equation. https://doi.org/10.1142.s021812741850133x.

11. T. Zhang and Z. Teng, *Nonlinear Analysis: Real World Applications* **9** 1409–1424 (2008). Global behaviors and permanence of SIRS epidemic model with time delay.

12. M. C. Mackay and I. Glass, *Science* **197**, 287–289 (1977). Oscillation and chaos in physiological control systems.

13. M. S. Gurney, S. P. Blythe, and R. M. Nisbee, *Nature* **287**, 17–21 (1980). Nicholson's blowflies revisited.

14. M. Wazewska-Czyzewska and A. Lasota, *Annals of the Polish Mathematical Society, Series III, Applied Mathematics* **17**, 23–40 (1976). Mathematical problems of the dynamics of the red blood cells system.

11.17   Fractional Differential Equations

Fractional differential equations are differential equations having one or more fractional derivatives [1]. Since, in general, few differential equations have exact solutions expressible as a finite combination of the elementary and special functions, discretization techniques must be formulated to compute numerical solutions [2].

As a quick glance toward what exactly are fractional derivatives [2], we present the following fast overview to the Grünwald-Letnikov derivative. The way this last statement is given clearly indicates that there are more than one procedure to introduce this topics.

The first derivative is defined as

$$f'(x) = \frac{df(x)}{dx} = \lim_{h \to 0} \frac{f(x+h) - f(x)}{h}. \tag{11.17.1}$$

Likewise, for the second derivative, we have

$$f''(x) = \frac{d^2 f(x)}{dx^2} \equiv \lim_{h \to 0} \frac{f'(x+h) - f'(x)}{h} \tag{11.17.2}$$

$$= \lim_{h \to 0} \frac{f(x+zh) - 2(f(x+h)) + f(x)}{h^2}. \tag{11.17.3}$$

For the general, n-th order derivative, where n is a positive integer, we have

$$f^n(x) = \frac{d^n(x)}{dx^n}$$

$$= \lim_{h \to 0} \left(\frac{1}{h^n}\right) \sum_{m=0}^{n} (-)^m \binom{n}{m} f[x + (n-m)h]. \tag{11.17.4}$$

Note that finite difference approximations for $f^{(n)}(x)$ can be derived by not taking the indicated limit on h to zero.

For non-integer n, the so-called Grünwald-Letnikov derivative is defined as follows

$$D^p f(x) \equiv \lim_{h \to 0} \left(\frac{1}{h^p}\right) \sum_{m=0}^{\infty} (-1)^m \binom{p}{m} f[x + p - m)h]. \tag{11.17.5}$$

If we define

$$\Delta_h^p f(x) = \sum_{m=0}^{\infty} (-)^m \binom{p}{m} f[x + (p-m)h], \tag{11.17.6}$$

then

$$D^p f(x) = \lim_{h \to 0} \left(\frac{1}{h^p}\right) \Delta_h^p f(x). \tag{11.17.7}$$

The references contain detailed discussions and numerical simulations of the following topics related to using the NSFD methodology to construct dynamical consistent discretizations for the relevant fractional differential equations:

- fluid dynamics [3]
- the Brusselater system [4]

- Hodgkin-Huxley model [5]
- a nonlinear wave equation [6]
- HIV-1 infection of $CD4^+$ T-cells [7]
- hyperbolic PDE's [8]
- SI and SIR epidemic models [9]
- class of haotic systems [10]
- neutron point kinetics [11]
- variable-order fractional differential equations [12]
- Navier-Stokes equation [13]
- salmonella transmission model [14]
- Swift-Hohenberg model [15]

References

1. K. S. Miller and B. Ross, *An Introduction to the Fractional Calculus and Fractional Differential Equations* (Wiley-Interscience, New York, 1993).

2. C. Lubich, *SIAM Journal of Mathematical Analysis* **17**, 704–719 (1986). Discretized fractional calculus.

3. K. Moaddy, S. Momani and I. Hashim, *Computers and Mathematics with Applications* **61**, 1209–1216 (2011). The non-standard finite difference scheme for linear fractional PDEs in fluid mechanics.

4. M. Y. Ongun, D. Arslan and R. Garrappa, *Advances in Difference Equations* **2013**, NSFD schemes for a fractional-order Brusselater system.
(http://www.advancesindifferenceequations.com/content/2013/1/102).

5. A. M. Nagy and N. H. Sweilam, *Physics Letters* **A378**, 1980–1984 (2014). An efficient method for solving fractional Hodgkin-Huxley model.

6. N. H. Sweilam and T. A. Assiri, *Progress in Fractional Differentiation and Applications* **1**, 269–280 (2015). Numerical scheme for solving the space-time variable order nonlinear fractional wave equation.

7. S. Zibaei and M. Namjoo, *Iranian Journal of Mathematical Chemistry* **6**, 169–184 (2015). A NSFD scheme for solving fractional-order model of HIV-1 inflection of $CD4^+$ T-cells.

8. N. H. Sweilam and T. A. Assiri, *Journal of Fractional Calculus and Applications* **7**, 46–60 (2016). NSFD schemes for solving fractional order byperbolic PDEs with Riesz fractional derivative.

9. A. J. Arenas, G. Gonzáldez-Parra and B. M. Chen-Charpentier, *Mathematics and Computers in Simulation* **121**, 48–63 (2016). Construction of NSFD schemes for the SI and SIR epidemic models of fractional order.

10. M. Hajipour, A. Jajarmi and D. Baleanu, *Journal of Computational and Nonlinear Dynamics* **13** (2), 021013 (December 07, 2017) (9 pages). Paper No:CND-16-1533; doi: 10.1115/1.4038444. An efficient NSFD scheme for a class of fractional chaotic systems.

11. Y. M. Hamada and M. G. Brikaa, *Annuals of Nuclear Energy* **102**, 359–367 (2017). NSFD schemes for numerical solution of the fractional neutron point kinetics equations.

12. A. M. Nagy, *Differential Equations and Dynamical Systems* (2017). https://doi.org/10.1007/s1259-017-0378-2. NSFD schemes for solving variable-order fractional differential equations.

13. K. Sayevand, T. Machado and V. Moradi, *Computers and Mathematics with Applications* (2018). A new NSFD method for analyzing the fractional Navier-Stokes equation. https://doi.org/10.1016/j.camwa.2018.12.016.

14. N. W. Sweilam, A. M. Nagy and L. E. Elfahri, *Journal of Fractional Calculus and Applications* **10**, 197–212 (2019). NSFD scheme for the fractional order salmonella transmission model.

15. W. K. Zahra, S. M. Eikholy and M. Fahmy, *Applied Mathematics and Computation* **343**, 372–387 (2019). Rational spline-NSFD scheme for the solution of time-fractional Swift-Hohenberg equation.

11.18   Summary

This chapter includes a broad range of systems modeled by various types of differential equations: ordinary and partial, delay, and fractional. In all cases, evidence was presented to demonstrate the usefulness, and value of applying the nonstandard finite difference (NSFD) methodology to construct dynamical consistent discretization schemes for the computation of numerical solutions to the relevant equations. We look forward to extensions of this work and the generalization of the NSFD methodology to other classes of differential equations such as those modelling stochastic processes.

Appendix A

Difference Equations

A.1 Linear Equations

Then n-th order linear inhomogeneous difference equation with constant coefficients has the form

$$y_{k+n} + a_1 y_{k+n-1} + a_2 y_{k+n-2} + \cdots + a_n y_k = R_k, \qquad (A.1.1)$$

where the $\{a_i\}$ are given constants, with $a_n \neq 0$, and R_k is a known function of k. If $R_k = 0$, then Eq. (A.1.1) is an n-th order homogeneous difference equation, i.e.,

$$y_{k+n} + a_1 y_{k+n-1} + a_2 y_{k+1n-2} + \cdots + a_n y_k = 0. \qquad (A.1.2)$$

The general solution to the homogeneous equation consists of a linear combination of n linearly independent functions $\{y_k^{(i)}\}$:

$$y_k^{(H)} = c_1 y_k^{(1)} + c_2 y_k^{(2)} + \cdots + c_n y_k^{(n)}, \qquad (A.1.3)$$

where the $\{c_i\}$ are n arbitrary constants. The n linearly independent functions $\{y_k^{(i)}\}$ are determined as follows [1]:

(i) First, construct the *characteristic equation* associated with Eq. (A.1.2); it is given by the expression

$$r^n + a_1 r^{n-1} + a_2 r^{n-2} + \cdots + a_n = 0. \qquad (A.1.4)$$

(ii) Denote the n roots of the characteristic equation by $\{r_i\}$.

(iii) Then n linearly independent functions $\{y_k^{(i)}\}$ are

$$y_k^{(i)} = (r_i)^k, \quad i = (1, 2, \ldots, n). \qquad (A.1.5)$$

Consequently, the general solution to Eq. (A.1.2) is

$$y_k^{(H)} = c_1 (r_1)^k + c_2 (r_2)^k + \cdots + c_n (r_n)^k. \qquad (A.1.6)$$

This result assumes that all the roots of the characteristic equation are distinct.

(iv) If root r_i occurs with a multiplicity $m \leq n$, then its contribution to the homogeneous solution takes the form

$$y_k^{(i)} = (A_1 + A_2 k + \cdots + A_m k^{m-1})(r_i)^k. \tag{A.1.7}$$

If $R_k \neq 0$, then the solution to the inhomogeneous Eq. (A.1.1) is a sum of the homogeneous solution $y_k^{(H)}$ and a particular solution to Eq. (A.1.1), i.e.,

$$y_k = y_k^{(H)} + y_k^{(P)}. \tag{A.1.8}$$

If R_k is a linear combination of various products of the terms

$$a^k, \quad e^{bk}, \quad \sin(ck), \quad \cos(ck), \quad k^\ell, \tag{A.1.9}$$

where (a, b, c) are constants and ℓ is a non-negative integer, then rules exist for constructing particular solutions [1]. When solutions to linear inhomogeneous equations are cited in this book, the particular solutions can usually be determined by inspection.

A.2 Riccati Equations

The following nonlinear, first-order difference equation

$$P y_{k+1} y_k + Q y_{k+1} + R y_k = S, \tag{A.2.1}$$

where (P, Q, R, S) are constants, appears in a number of places in the deliberations of this book. This equation is a special case of Riccati difference equation [1]. This equation can be solved exactly by first dividing by P and shifting the index k to give

$$y_k y_{k-1} + A y_k + B y_{k-1} = C, \tag{A.2.2}$$

where

$$A = \frac{Q}{P}, \quad B = \frac{R}{P}, \quad C = \frac{S}{P}. \tag{A.2.3}$$

The nonlinear transformation

$$y_k = \frac{x_k - B x_{k+1}}{x_{k+1}}, \tag{A.2.4}$$

reduces Eq. (A.2.2) to the following linear, second-order equation for x_k:

$$(AB + C) x_{k+1} - (A - B) x_k - x_{k-1} = 0. \tag{A.2.5}$$

This equation can be solved by the method of Section A.1 and consequently, the general solution to Eq. (A.2.1) can be found.

Note that if $S = 0$, Eq. (A.2.1) becomes

$$Py_{k+1}y_k + Qy_{k+1} + Ry_k = 0 \tag{A.2.6}$$

and the substitution

$$y_k = \frac{1}{x_k}, \tag{A.2.7}$$

gives the first-order, linear difference equation

$$Rx_{k+1} + Qx_k + P = 0, \tag{A.2.8}$$

which can be easily solved.

A.3   Separation-of-Variables

Many partial difference equations with constant coefficients can be solved to obtain special solutions by the method of separation-of-variables [1],

Let $z(k, \ell)$ denote a function of the discrete (integer) variables (k, ℓ). Now define the two shift operators, E_1 and E_2, as follows

$$(E_1)^m z(k, \ell) = z(k + m, \ell), \tag{A.3.1}$$

$$(E_2)^m z(k, \ell) = z(k, \ell + m), \tag{A.3.2}$$

where m is an integer. Now consider the linear partial difference equation

$$\psi(E_1, E_2, k, \ell) z(k, \ell) = 0, \tag{A.3.3}$$

where the operator ψ is a polynomial function of E_1 and E_2. The basic principle of the method of separation of variables is to assume that $z(k, \ell)$ can be written as

$$z(k, \ell) = C(k)D(\ell). \tag{A.3.4}$$

Assume further that when this form is substituted into Eq. (A.3.3), an equation having the structure

$$\frac{f_1(E_1, k)C(k)}{f_2(E_1, k)C(k)} = \frac{g_1(E_2, \ell)D(\ell)}{g_2(E_2, \ell)D(\ell)}, \tag{A.3.5}$$

is obtained. Under these conditions, $C(k)$ and $D(\ell)$ satisfy the ordinary difference equations

$$f_1(E_1, k)C(k) = \alpha f_2(E_1, k)C(k) \tag{A.3.6}$$

$$g_1(E_2, \ell)D(\ell) = \alpha g_2(E_2, \ell)D(\ell), \tag{A.3.7}$$

where α is the arbitrary separation constant. The solutions to these equations will depend on α, i.e., $C(k, \alpha)$ and $D(\ell, \alpha)$. Therefore, the special solution $z(k, \ell)$ given by Eq. (A.3.4) will also depend on α. Since the original partial difference equation is linear, a general solution can be obtained by "summing" over α, i.e.,

$$z(k, \ell) = \sum \!\!\!\!\!\!\int z(k, \ell, \alpha)d\alpha, \tag{A.3.8}$$

where

$$z(k, \ell, \alpha) = C(k, \alpha)D(\ell, \alpha). \tag{A.3.9}$$

Reference

1. R. E. Mickens, *Difference Equations: Theory and Applications* (Van Nostrand Reinhold, New York, 1990). See Sections 4.2, 4.5, 5.3 and 6.3.

Appendix B

Linear Stability Analysis

B.1 Ordinary Differential Equations

Consider the scalar ordinary differential equation

$$\frac{dy}{dt} = f(y). \tag{B.1.1}$$

Assume that

$$f(\bar{y}) = 0, \tag{B.1.2}$$

has m simple zeros, where m may be unbounded. The solutions of Eq. (B.1.2) are fixed-points of the differential equation and correspond to constant solutions.

The linear stability properties of the fixed-points are determined by investigating the behavior of small perturbations about a given fixed-point [1, 2]. Consider the i-th fixed-point, $\bar{y}^{(i)}$. The perturbed trajectory takes the form

$$y(t) = \bar{y}^{(i)} + \epsilon(t), \tag{B.1.3}$$

where

$$|\epsilon(t)| \ll |\bar{y}^{(i)}|. \tag{B.1.4}$$

Substitution of Eq. (B.1.3) into Eq. (B.1.1) gives

$$\frac{d\epsilon}{dt} = f|\bar{y}^{(i)}| + R_i\epsilon + O(\epsilon^2), \tag{B.1.5}$$

where

$$R_i = \left.\frac{df}{dy}\right|_{y=\bar{y}^{(i)}}. \tag{B.1.6}$$

The linear stability equation is given by the linear terms of Eq. (B.1.5), i.e.,

$$\frac{d\epsilon}{dt} = R_i\epsilon. \tag{B.1.7}$$

The solution of this equation is

$$\epsilon(t) = \epsilon_0 e^{R_i t}. \tag{B.1.8}$$

The fixed-point, $y^{(t)} = \bar{y}^{(i)}$, is said to be linearly stable if $R_i < 0$, and linearly unstable if $R_i > 0$.

These results can be easily generalized to the case of coupled first-order differential equations [1, 2].

B.2   Ordinary Difference Equations

The fixed-points of the first-order difference equation

$$y_{k+1} = F(y_k), \tag{B.2.1}$$

are the solutions to

$$\bar{y} = F(\bar{y}). \tag{B.2.2}$$

Assume that all the zeros of

$$G(\bar{y}) = F(\bar{y}) - \bar{y} = 0, \tag{B.2.3}$$

are simple and denote the n fixed-points by $\{\bar{y}^{(j)}\}$, $j = (1, 2, \ldots, n)$. The perturbed solution about the particular fixed-point

$$y_k = \bar{y}^{(j)} \tag{B.2.4}$$

can be written as

$$y_k = \bar{y}^{(j)} + \epsilon_k, \tag{B.2.5}$$

where

$$|\epsilon_k| \ll |\bar{y}^{(j)}|. \tag{B.2.6}$$

Substitution of Eq. (B.2.5) into Eq. (B.2.1) and retaining only linear terms in ϵ_k gives

$$\epsilon_{k+1} = R_j \epsilon_k \tag{B.2.7}$$

where

$$R_j = \frac{dF}{dy}\bigg|_{y=\bar{y}^{(j)}}. \tag{B.2.8}$$

Equation (B.2.7) has the solution

$$e_k = e_0 (R_j)^k. \tag{B.2.9}$$

Thus, the j-th fixed-point of Eq. (B.2.1) is said to be linearly stable if $|R_j| < 1$ and linearly unstable for $|R_j| > 0$.

Additional details and the generalization to higher-order systems of difference equations are discussed in references [3, 4, 5, 6].

References

1. L. Cesari, *Asymptotic Behavior and Stability Problems in Ordinary Equations* (Academic Press, New York, 2nd edition, 1963).
2. D. A. Sanchez, *Ordinary Differential Equations and Stability Theory* (W. H. Freeman, San Francisco, 1968).
3. R. E. Mickens, *Difference Equations: Theory and Applications* (Van Nostrand Reinhold, New York, 1990). See Section 7.4.
4. J. T. Sandefur, *Discrete Dynamical Systems: Theory and Applications* (Clarendon Press, Oxford, 1990). See Chapter 4.
5. E. I. Jury, *IEEE Transactions on Automatic Control* **AC-16**, 233–240 (1971). The inners approach to some problems of system theory.
6. E. R. Lewis, *Network Models in Population Biology* (Springer-Verlag, New York, 1977).

Appendix C

Discrete WKB Method

In Section 8.2, we have made use of a discrete version of the WKB method [1] to calculate the asymptotic behavior of difference equations having the form

$$y_{k+1} + y_{k-1} = 2\sigma_k y_k, \tag{C.1}$$

where σ_k has the asymptotic representation

$$\sigma_k = A_0 + \frac{A_1}{k} + \frac{A_2}{k^2} + \frac{A_3}{k^3} + O\left(\frac{1}{k^4}\right), \tag{C.2}$$

with

$$|A_0| \neq 1. \tag{C.3}$$

The asymptotic behavior of y_k is given by the expression [2]

$$y_k = k^\theta e^{B_0 k} \left[1 + \frac{B_1}{k} + \frac{B_2}{k^2} + \frac{B_3}{k^3} + O\left(\frac{1}{k^4}\right)\right], \tag{C.4}$$

where $(\theta, B_0, B_1, B_2, B_3)$ may be complex valued and are to be determined as functions of (A_0, A_1, A_2, A_3).

If the form of Eq. (C.4) is substituted in to Eq. (C.1) and use is made of the relation

$$(k \pm 1)^m = k^m \left(1 \pm \frac{1}{k}\right)^m = k^m \left\{1 \pm \frac{m}{k} \left[\frac{m(m-1)}{2}\right] \frac{1}{k^3}\right.$$

$$\left. \pm \left[\frac{m(m-1)(m-2)}{6}\right] \frac{1}{k^3} + O\left(\frac{1}{k^4}\right)\right\}, \tag{C.5}$$

then the setting to zero of the coefficients of the terms

$$k^\theta e^{B_0 k} \left(\frac{1}{k^m}\right), \quad m = (0, 1, 2, 3),$$

305

gives a set of equations that can be solved for (θ, B_0, B_1, B_2). They are

$$B_0 = \pm \cosh^{-1}(A_0) = \text{Ln}\left(A_0 \pm \sqrt{A_0^2 - 1}\right), \tag{C.6}$$

$$\theta = \frac{A_1}{\sinh(B_0)}, \tag{C.7}$$

$$B_1 = -\frac{A_2}{\sinh(B_0)} + \left[\frac{\theta(\theta - 1)}{2}\right]\tanh(B_0), \tag{C.8}$$

$$B_2 = \frac{[(1 - \theta)\cosh(B_0) + \theta(\theta - 1)/2 - A_2]B_1}{2\sinh(B_0)}$$
$$- \frac{A_3}{2\sinh(B_0)} + \frac{\theta(\theta - 1)(\theta - 2)}{12}. \tag{C.9}$$

The above relations can be rewritten without the use of hyperbolic functions by making the following replacements:

$$\cosh(B_0) = A_0, \tag{C.10}$$

$$\sinh(B_0) = \pm\sqrt{A_0^2 - 1}. \tag{C.11}$$

References

1. J. D. Murray, *Asymptotic Analysis* (Springer-Verlag, New York, 1984).
2. J. Wimp, *Computation with Recurrence Relations* (Pitman, Boston, 1984). See Appendix B.
3. R. E. Mickens and I. Ramadhani, WKB procedures for Schrödinger type difference equations, in *Proceedings of the First World Congress of Nonlinear Analysts* (Tampa, FL; August 19–26, 1992), pp. 3907–3912.

Bibliography

DIFFERENCE EQUATIONS

R. P. Agarwal, *Difference Equations and Inequalities* (Marcel Dekker, New York, 1992).

P. M. Batchelder, *An Introduction to Linear Difference Equations* (Harvard University Press, Cambridge, 1927).

G. Boole, *Calculus of Finite Differences* (Chelsea, New York, 4th edition, 1958).

L. Brand, *Differential and Difference Equations* (Wiley, New York, 1966).

F. Chorlton, *Differential and Difference Equations* (Van Nostrand, London, 1965).

E. J. Cogan and R. Z. Norman, *Handbook of Calculus, Difference and Differential Equations* (Prentice-Hall; Englewood Cliffs, NJ; 1958).

S. Elaydi, *An Introduction to Difference Equations*, 3rd Edition (Springer-Verlag, New York, 2005).

T. Fort, *Finite Differences and Difference Equations in the Real Domain* (Clarendon Press, Oxford, 1948).

P. E. Hydon, *Difference Equations by Differential Equations* (Cambridge University Press, New York, 2014).

C. Jordan, *Calculus of Finite Differences* (Chelsea, New York, 3rd edition, 1965).

E. I. Jury, *Theory and Applications of the z-Transform Method* (Wiley, New York, 1964).

W. G. Kelley and A. C. Peterson, *Difference Equations: An Introduction with Applications* (Academic Press, Boston, 1991).

H. Levy and F. Lessman, *Finite Difference Equations* (Macmillan, New York, 1961).

R. Li, Z. Chen, and W. Wu, *Generalized Difference Methods for Differential Equations* (Marcel Dekker, New York, 2000).

R. E. Mickens, *Difference Equations: Theory and Applications* (Van Nostrand Reinhold, New York, 1990).

K. S. Miller, *An Introduction to the Calculus of Finite Differences and Equations* (Holt, New York, 1960).

K. S. Miller, *Linear Difference Equations* (W. A. Benjamin, New York, 1968).

L. M. Milne-Thomson, *The Calculus of Finite Differences* (Macmillan, London, 1960).

C. H. Richardson, *An Introduction to the Calculus of Finite Differences* (Van Nostrand, New York, 1954).

J. T. Sandefur, *Discrete Dynamical Systems: Theory and Applications* (Clarendon Press, Oxford, 1990).

H. Sedaghat, *Nonlinear Difference Equations: Theory and Applications to Social Science Models* (Springer, New York, 1999).

A. N. Sharkovsky, Yu. L. Maistrenko and E. Yu. Romanenko, *Difference Equations and Their Applications* (Kluwer Academic, Dodrecht, 1993).

M. R. Spiegel, *Calculus of Finite Differences and Difference Equations* (McGraw-Hill, New York, 1971).

NUMERICAL ANALYSIS: THEORY AND APPLICATIONS

F. S. Acton, *Numerical Methods that Work* (Mathematical Association of America; Washington, DC; 1990).

M. B. Allen III, I. Herrerra and G. F. Pinder, *Numerical Modeling in Science and Engineering* (Wiley-Interscience, New York, 1988).

J. B. Botha and G. F. Pinder, *Fundamental Concepts in the Numerical Solutions of Differential Equations* (Wiley-Interscience, New York, 1983).

J. C. Butcher, *The Numerical Analysis of Ordinary Differential Equations: Runge-Kutta and General Linear Methods* (Wiley-Interscience, Chichester, 1987).

K. Eriksson, D. Estep, P. Hansbo, and C. Johnson, *Computational Differential Equations* (Cambridge University Press, New York, 1996).

E. Gekeler, *Discretization Methods for Stable Initial Value Problems* (Springer-Verlag, Berlin, 1984).

D. Greenspan, *Discrete Models* (Addison-Wesley; Reading, MA; 1973).

D. Greenspan and V. Casulli, *Numerical Analysis for Applied Mathematics, Science, and Engineering* (Addison-Wesley; Redwood City, CA; 1988).

F. B. Hildebrand, *Finite-Difference Equations and Simulations* (Prentice-Hall; Englewood Cliffs, NJ; 1968).

M. Holt, *Numerical Methods in Fluid Dynamics* (Springer-Verlag, Berlin, 1984).

A. Eserles, *A First Course in the Numerical Analysis of Differential Equations* (Cambridge University Press, New York, 1996).

M. K. Jain, *Numerical Solution of Differential Equations* (Wiley, New York, 2nd edition, 1984).

L. Lapidus and G. F. Pinder, *Numerical Solution of Partial Differential Equations in Science and Engineering* (Wiley-Interscience, New York, 1982).

V. Lakshmikantham and D. Trigiante, *Theory of Difference Equations: Numerical Methods and Applications* (Academic Press, Boston, 1988).

W. J. Lick, *Difference Equations from Differential Equations* (Springer-Verlag, Berlin, 1989).

W. E. Milne, *Numerical Solution of Differential Equations* (Dover, New York, 1970).

J. M. Ortega and W. G. Poole, Jr., *An Introduction to Numerical Differential Equations* (Pitman; Marshfield, MA; 1981).

D. Potter, *Computational Physics* (Wiley-Interscience, Chichester, 1973).

A. Quarteroni, R. Sacco, and F. Saleri, *Numerical Mathematics* (Springer, New York, 2000).

R. D. Richtmyer and K. W. Morton, *Difference Methods for Initial-Value Problems* (Interscience, New York, 1967).

P. J. Roache, *Computational Fluid Dynamics* (Hermosa Publishers; Albuquerque, NM; 1976).

G. D. Smith, *Numerical Solution of Partial Differential Equations: Finite Difference Methods* (Clarendon Press, Oxford, 2nd edition, 1978).

H. J. Stetter, *Analysis of Discretization Methods for Ordinary Differential Equations* (Springer-Verlag, Berlin, 1973).

J. C. Strikwerda, *Finite Difference Schemes and Partial Differential Equations* (Wadsworth and Brooks/Cole; Belmont, CA; 1989).

R. Vichnevetsky and J. B. Bowles, *Fourier Analysis of Numerical Approximations of Hyperbolic Equations* (Society for Industrial and Applied Mathematics, Philadelphia, 1982).

J. Wimp, *Computation with Recurrence Relations* (Pitman, Boston, 1984).

GENERAL PUBLICATIONS ON NONSTANDARD SCHEMES

S. Lemeshevsky, P. Matus, and D. Poliakov (editors), *Exact Finite-Difference Schemes* (De Gruyter, Berlin, 2016).

R. E. Mickens (editor), *Applications of Nonstandard Finite Difference Schemes* (World Scientific, Singapore, 2000).

R. E. Mickens (editor), *Advances in the Applications of Nonstandard Finite Difference Schemes* (World Scientific, Singapore, 2005).

K. C. Patidar, "Nonstandard finite difference methods: Recent trends and further developments," *Journal of Difference Equations and Applications* **22**, 817 (2016).

Index